BIOLOGICAL CONTROL OF WEEDS WITH PLANT PATHOGENS

BIOLOGICAL CONTROL OF WEEDS WITH PLANT PATHOGENS

Edited by

R. CHARUDATTAN
University of Florida

H. L. WALKER
USDA–ARS, Southern Weed Science Laboratory

1807 1982

175 YEARS OF PUBLISHING

A Wiley-Interscience Publication
JOHN WILEY & SONS
New York   Chichester   Brisbane   Toronto   Singapore

Library of Congress Cataloging in Publication Data:
Main entry under title:
Biological control of weeds with plant pathogens.

   "A Wiley-Interscience publication."
   Includes index.
   1. Weed control—Biological control.   2. Weeds—
Diseases and pests.   3. Micro-organisms, Phyto-
pathogenic.   I. Charudattan, R.   II. Walker, H. L.

SB611.5.B56   632′.58   82-1879
ISBN 0-471-08598-7   AACR2

Printed in the United States of America

10   9   8   7   6   5   4   3   2   1

CONTRIBUTORS

SPENCER C. H. BARRETT, Department of Botany, University of Toronto, Toronto, Ontario, Canada

R. CHUCK BOWERS, Agricultural Division, The Upjohn Company, Kalamazoo, Michigan

JAMES M. CHANDLER, USDA–ARS, Southern Weed Science Laboratory, Stoneville, Mississippi

RAGHAVAN CHARUDATTAN, Plant Pathology Department, University of Florida, Gainesville, Florida

BRUCE W. CHURCHILL, Fermentation Research and Development, The Upjohn Company, Kalamazoo, Michigan

RICHARD E. CULLEN, Plant Pathology Department, University of Florida, Gainesville, Florida

WILLIAM E. DYER, Department of Plant and Soil Science, Montana State University, Bozeman, Montana

NABIH E. EL–GHOLL, Division of Plant Industry, Florida Department of Agriculture and Consumer Services, Gainesville, Florida

PETER K. FAY, Department of Plant and Soil Science, Montana State University, Bozeman, Montana

T. ED. FREEMAN, Plant Pathology Department, University of Florida, Gainesville, Florida

RICHARD T. HANLIN, Department of Plant Pathology, University of Georgia, Athens, Georgia

DAVID S. HERON, Plant Pathology Department, University of Florida, Gainesville, Florida

GORDON E. HOLCOMB, Department of Plant Pathology and Crop Physiology, Louisiana State University, Baton Rouge, Louisiana

CURTIS R. JACKSON, Georgia Experiment Station, Experiment, Georgia

M. ELIZABETH KANNWISCHER, Plant Pathology Department, University of Florida, Gainesville, Florida

DONALD S. KENNEY, Microbial Products Research, Abbott Research Center, Abbott Laboratories, Long Grove, Illinois

KURT J. LEONARD, USDA–ARS, Department of Plant Pathology, North Carolina State University, Raleigh, North Carolina

CHESTER G. McWHORTER, USDA–ARS, Southern Weed Science Laboratory, Stoneville, Mississippi

PAUL C. QUIMBY, JR., USDA–ARS, Southern Weed Science Laboratory, Stoneville, Mississippi

WILLIAM H. RIDINGS, Department of Plant Pathology and Physiology, Clemson University, Clemson, South Carolina

DAVID C. SANDS, Department of Plant Pathology, Montana State University, Bozeman, Montana

CALVIN L. SCHOULTIES, Bureau of Plant Pathology, Division of Plant Industry, Florida Department of Agriculture and Consumer Services, Gainesville, Florida

EUGENE L. SHARP, Department of Plant Pathology, Montana State University, Bozeman, Montana

ROBERT D. SHRUM, USDA–ARS, Plant Disease Research Laboratory, Frederick, Maryland

ROY J. SMITH, JR., USDA–ARS, Stuttgart, Arkansas

HARVEY W. SPURR, JR., USDA–ARS, Oxford Tobacco Laboratory, Oxford, North Carolina

DAVID O. TeBEEST, Department of Plant Pathology, University of Arkansas, Fayetteville, Arkansas

GEORGE E. TEMPLETON, Department of Plant Pathology, University of Arkansas, Fayetteville, Arkansas

SHERRY K. TURNER, Department of Plant and Soil Science, Montana State University, Bozeman, Montana

C. GERALD VAN DYKE, Botany Department, North Carolina State University, Raleigh, North Carolina

H. LYNN WALKER, USDA–ARS, Southern Weed Science Laboratory, Stoneville, Mississippi

ALAN K. WATSON, Plant Science Department, Macdonald Campus, McGill University, Ste. Anne De Bellevue, Quebec, Canada

SUZAN H. WOODHEAD, Microbial Products Research, Abbott Research Center, Abbott Laboratories, Long Grove, Illinois

A. DOUGLAS WORSHAM, Crop Science Department, North Carolina State University, Raleigh, North Carolina

PREFACE

Biological Control of Weeds with Plant Pathogens is the result of an international workshop organized by the members of the U.S. Department of Agriculture and State Agricultural Experiment Station Southern Regional Research Project S-136, "Biological Control of Weeds with Fungal Plant Pathogens," to critique research on the use of plant pathogens for weed control. The workshop, sponsored by the University of Arkansas and the U.S. Department of Agriculture, Science and Education Administration, Cooperative Research, was held on September 8–10, 1980 in Little Rock, Arkansas. Participants included weed scientists, plant pathologists, state and federal regulatory personnel, industrial scientists, and administrative personnel. The workshop consisted of invited papers by internationally recognized scientists and a poster session summarizing ongoing research in this area of biological weed control.

This book, which consolidates the information presented during the workshop, represents the first multidisciplinary treatment of this subject area. It provides a source of information for scientists, teachers, and students who are interested in the use of plant pathogens to control weeds. Although the future of this research area is promising, the biological weed control goals and strategies presented here are expected to complement existing weed management programs and should not be construed as a challenge to conventional weed control methods.

The status and strategies of conventional and biological weed control methods are discussed, with emphasis on the concepts, benefits, and problems concerning the use of plant pathogens as biological weed control agents. One chapter is devoted to examples of historically important plant diseases. Host-pathogen interactions are discussed with emphasis on the aspects of these interactions that are particularly relevant to biological weed control. Potential hazards and benefits that relate to the use of plant pathogens as biological weed control agents are discussed, as are the various governmental regulations that apply to plant pathogens used in the classical or microbial herbicide weed control strategies. Mass production and commercialization of plant pathogens are described as these relate to the integration of microbial herbicides into pest management systems. A chapter devoted to literature retrieval provides valuable information for establishing and updating bibliographies for biological control research. These topics are arranged in four sections: Introduction, Host-Parasite Interaction, Microbial Weed Control Agents, and Summary. The appendixes provide lists of scientific names of plant pathogens and arthropods, the common and scientific names of host plants, and the common and chemical names of pesticides cited in the book.

We wish to acknowledge the members of S-136 and all other participants for making the workshop successful. Much credit is due to the organizational efforts of George E. Templeton. David O. TeBeest organized the poster session at this workshop; the abstracts of these posters are presented in Appendix 1. Financial support for the workshop was provided by the U.S. Department of Agriculture, Cooperative Research. We are grateful to Donald K. Crawley, Richard E. Cullen, David S. Heron, Frank M. Hofmeister, Joseph A. Riley, and Robert N. Warrington for assistance with the final preparation of the book.

R. CHARUDATTAN
H. LYNN WALKER

Gainesville, Florida
Stoneville, Mississippi
March 1982

CONTENTS

III. MICROBIAL WEED CONTROL AGENTS

BIOLOGICAL CONTROL OF WEEDS WITH PLANT PATHOGENS

I

INTRODUCTION

OBJECTIVES

C. R. JACKSON
Georgia Experiment Station, Experiment, Georgia

The deliberate use of plant pathogens to achieve economic control of weeds is a new subject. The subject brings with it a welcomed attempt to integrate plant pathology, weed science, and plant physiology in their broadest senses. A glance at the ensuing chapters will confirm this and indicate the broad array of talented persons interested in this subject.

One objective of this book is to give insights into the research now under way and the accomplishments to this point. The successes that have been achieved or are imminent are clearly presented in the abstracts in Appendix 1. The abstracts depict research on the control of alligatorweed (see Appendix 2 for botanical names of plants cited in this book), Canada thistle, cocklebur, milkweed vine, morningglory, northern jointvetch, prickly sida, Russian knapweed, spurred anoda, and waterhyacinth. The control agents are fungi in most cases; we must give more thought to the use of viruses and bacteria. Also discussed in the abstracts are the methods for isolation and identification of weed-infecting fungi and key decisions that should mark the progression of research in this area.

The use of native pathogens involves many knotty problems, and these problems are greatly amplified when we consider the use of exotic pathogens. A principal objective of this book is to discuss examples of weed control with plant pathogens, the constraints to successful use of pathogens, the importance of genetic variation in weeds and pathogens, and problems of establishing a pathogen-weed relationship of sufficient magnitude to obtain control. Perhaps we will be stimulated to consider the prospects for soil-borne pathogens, obligate parasites, and integration of all of these control tools with other pest management procedures to achieve weed control.

Our neat and proven research models may show how to control weeds with pathogens, but the knowledge gained through this research will not serve agriculture well unless these concepts and ideas are put into commercial use. For commercial development of these bioherbicides, we must rely on industry interests and initiatives that must operate within economic rigors and harmonize with partially unknown state and federal regulations. Therefore, the subjects of commercialization and regulation have been included in the book so that we may be aware of current industry and government outlook on this area.

The task of scientists is to determine the most promising courses of action on the basis of available information. This book provides pertinent information regarding the state of the art for biological weed control with plant pathogens. We hope that more scientists will be stimulated to work in this area of biological control. Perhaps equally important, we hope that this book will evoke a movement toward focusing ideas and scientists to work together to raise this broad subject to national and international prominence as a science issue. The information contained in this book is the proper nucleus for such an undertaking.

CONVENTIONAL WEED CONTROL TECHNOLOGY

C. G. McWHORTER AND J. M. CHANDLER

Southern Weed Science Laboratory,
Science and Education Administration, Agricultural Research,
U.S. Department of Agriculture, Stoneville, Mississippi

Significant advances have been made in weed control technology during the last 20 years, but control techniques have not been developed for many specific weeds. There are over 300,000 species of plants throughout the world, but only about 30,000 of these are weeds. About 1800 of the weeds cause serious economic losses in crop production, and about 300 weed species are serious in cultivated crops throughout the world (1). Most cultivated plants are plagued by 10–30 weed species that must be controlled to avoid yield reductions.

Holm et al. (2) listed about 200 weed species that cause 95% of weed problems for humans. They list 80 species as primary weeds and 120 species as secondary weeds. Seventy percent of the world's worst weeds are present in the United States. Sixty-seven percent of those present in the United States are broadleaf weeds, 24% are grasses, and 7% are sedges (2).

Even with the use of our most advanced weed control technology, there are many individual weed species that cause crop losses amounting to billions of dollars annually. In 1965 it was estimated that crop losses

Cooperative investigations of the Southern Weed Science Laboratory, Science and Education Administration, Agricultural Research, U.S. Department of Agriculture, and the Delta Branch, Mississippi Agriculture and Forestry Experiment Station.

caused by wild oat alone and the cost of its control amounted to more than $300 million annually (3,4). Others, such as quackgrass and johnsongrass, have been impossible to control selectively in small grains (5). Perennial weeds such as wild garlic, wild onion, field bindweed, Canada thistle, and perennial sowthistle are currently held in check by postemergence applications of (2,4-dichlorophenoxy)acetic acid (2,4-D) (5). Even with 2,4-D, continued treatment is needed for several years to reduce weed infestations significantly. No herbicides are available for the control of many annual grasses in small grains, including jointed goatgrass, weedy bromes, hairy chess, Japanese brome, cheat, foxtails, and barnyardgrass. Perennial broadleaf weed problems appear to be increasing in severity throughout the United States in many crops (5).

In addition to the many troublesome weeds for which there are no adequate controls, new weeds are appearing that create major weed control problems (6-9). The new weeds most often appear in a monoculture situation after continuous use of specific herbicide programs. Technology has not been available to predict these new problems that probably arise because there are so many different kinds of weeds with varying periods of germination and highly divergent life cycles. The management of diverse weed populations requires an integrated systems approach that employs chemical, cultural, mechanical, biological, ecological, and bioenvironmental methods. Unfortunately, weed research has not been successful in providing control to handle such complicated weed management programs in all crops.

METHODS OF WEED CONTROL

There are seven general methods of weed control: crop competition, crop rotation, biological, fire, mechanical, hand labor, and chemical. Usually, the best and most economical way of controlling weeds is a combination of these methods. Many farmers have used combined methods for years without realizing that they were utilizing integrated weed management systems.

CROP COMPETITION

Crop competition is often one of the most economical weed control methods available to farmers. For example, most soybean producers know that an inadequate stand of soybean in early season is an invitation to increased weed competition. Also, many soybean cultivars are more competitive than others with weeds. Therefore, farmers may select a soybean variety based on the severity of the weed problem. For example, the Mississippi State Weed Science Committee recommends certain soybean cultivars for use in those situations where it is known in advance that severe weed competition may occur (10).

Adequate stands of cotton or soybean can drastically reduce the development of a weed such as spurred anoda (11); the leaf area index (LAI) of spurred anoda can be reduced 88% by soybean and 73% by cotton (Fig. 1). In this case, when competing with cotton or soybean a spurred anoda LAI of 1.0 occurred at 7 and 10 weeks, respectively, compared to an LAI of about 3.0 in noncompeting stand of spurred anoda. Crop competition does not often eliminate a weed problem, but it is an important constituent in some of the most successful integrated weed management systems.

CROP ROTATION

The rotation of crops is often an efficient way to minimize weed competition. The best rotations for weed control usually include highly competitive crops in each part of the rotation involving both summer row crops and winter or spring grain crops. Spurred anoda, for example, is

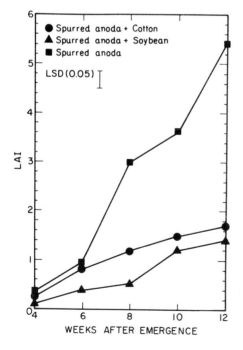

FIGURE 1. Effect of cotton or soybean competition on the leaf area index (LAI) of spurred anoda at 80 plants per 12 m of row.

highly competitive with cotton, but it is far less competitive with soybean (11). In addition, spurred anoda is controlled much more efficiently in soybean with herbicides such as bentazon than by the herbicides used in cotton for control of this weed (11). Thus the rotation of crops also permits herbicide rotations, and this has become increasingly important in developing more effective integrated weed management systems. Many annual and perennial weeds can be effectively and economically controlled utilizing a crop-herbicide rotation. For example, the biomass of johnsongrass can be decreased drastically and corn yields increased by using a corn-cotton-cotton-corn sequence as compared with continuous corn (Table 1). In the corn-cotton sequence the level of herbicide input has been observed to significantly influence the level of johnsongrass infestation and corn yield during the fourth year (12).

BIOLOGICAL CONTROL

Historically, biological control has worked best on large infestations of a single weed species. These situations usually occur in rangelands or in waterways. Also, biological control has been most successful where a weed was introduced into a new area freed from its natural enemies.

Examples of successful biological weed control include the control of species of *Opuntia* cactus in Australia by larvae of a moth imported from Argentina and the control of St. Johnswort in the western United States by leaf-eating beetles. Unfortunately, biological weed control has not developed to the point that it has had any appreciable impact on the production of agronomic crops in the United States. Biological weed control agents are needed in cropping situations in the United States to permit development and successful use of realistic integrated weed management systems. Comparatively little research has been devoted to the discovery of biological weed control agents for use in crop production, and a greatly expanded effort—in both monies and scientific effort—is needed.

FIRE

Fire has been used for many decades to control weeds in forests, on ditchbanks, on roadsides, and other waste areas. Fire (flame cultivation) has also been used to control weeds in alfalfa and cotton. The most successful way to employ flame cultivation in cotton is to use flame burners when weeds are small and to reapply this control technique at regular intervals. However, the successful use of flame in cotton production and in other crops requires a high level of management.

Flame cultivation is only rarely used at present. The availability of se-

TABLE 1. Relation of Corn Yield to Biomass of Johnsongrass Present at the End of the Fourth Year of Corn Grown in Sequences of Corn-Cotton-Cotton-Corn versus Continuous Corn

Herbicide Treatment by Crop and Crop Sequence[a]		Numbers of Johnsongrass[b] (Stems/m^2)	Dry Weight Johnsongrass[b] (g/m^2)	Corn Yield[b] (kg/ha)
Corn (kg/ha)	Cotton (kg/ha)			
Corn-Cotton-Cotton-Corn				
Hand-weeded control	Hand-weeded control	13 b	10 c	5660 a
Cultivated control	Cultivated control	26 b	160 bc	1390 d
Atrazine; 2.24, PE	Trifluralin; 0.56, PI	6 b	106 c	3960 bc
Atrazine; 2.24, PE	Trifluralin; 0.56, PI + fluometuron; 2.24, PE	3 b	0 c	4670 abc
Atrazine; 2.24, PE + linuron; 0.84; POE DIR	Fluometuron; 2.24, PE	12 b	91 c	4560 abc
Atrazine; 2.24, PE + linuron; 0.84, POE DIR	Trifluralin; 0.56, PI + fluometuron; 2.24, PE + MSMA; 2.24, POE DIR	4 b	5 c	5580 a
Continuous Corn				
Hand-weeded control	—	30 b	88 c	3870 bc
Cultivated control	—	88 a	347 bc	670 d
Atrazine; 2.24, PE	—	61 a	377 a	550 d
Linuron; 0.84, POE DIR	—	86 a	432 a	590 d
Atrazine; 2.24, PE + linuron; 0.84, POE DIR	—	72 a	511 a	1670 d

Source: Adapted from Ref. 12.

[a]Herbicide application methods are designated by the following symbols: MSMA, monosodium methanearsonate; PE, preemergence; PI, preplant incorporated; POE DIR, postemergence directed. See Appendix 4 for chemical names of herbicides cited here and elsewhere in the book.

[b]Means within columns followed by the same letter do not differ at the 5% level of probability according to Duncan's multiple range test.

lective herbicides and the increased cost of fuel have contributed to the reduced use of flame as a weed control agent. It is unlikely that flame will be an important constituent of integrated weed management systems in the future.

MECHANICAL CONTROL

Cultivation has been employed for weed control for centuries in most crops throughout the world. In fact, the method by which many crops are produced has been dictated by the need for cultivation. The spacing of rows 0.5–1.0 m apart in many row crops was fixed by the need to till the soil for weed control.

The cultivation of row crops in the United States increased from about 56 million ha in 1965 to about 68 million ha in 1973 (13). The number of cultivations normally required for effective weed control varies from two to five, depending on the crop. It has been estimated that about 50% of all tillage in the United States is needed specifically for weed control. Wiese and Chandler (13) reported that cultivation in the United States was accomplished with a fleet of 4.5 million tractors that used up to 75 billion liters of crude oil annually. There has been emphasis in recent years on reduction of the number of cultivations, not only to conserve energy, but also to control soil erosion. Reduced or minimum tillage systems have worked best in row crops and in some areas in the United States where weeds could be controlled only with herbicides. In other areas of the United States, such as the Southeast, cultivation is still an important part of integrated weed management systems. There are many advantages of reduced tillage crop production systems, but minimum tillage will not become the conventional method of crop production, especially in the Southeast, until more effective weed control programs are provided.

HAND LABOR

Hand labor was a conventional method of weed control in many row crops in the United States until the 1950s. There has been a drastic reduction in the use of unskilled hoe labor for weed control in row crops during the last 30 years, as the farm work force decreased and the price of hoe labor greatly increased. Hoe labor cost throughout much of the Southeast in the 1950s was about $0.35/hour, and this cost increased to about $3.65/hour in 1980. Labor for hand hoeing was also sharply decreased after 1963 when the Labor Program with Mexico terminated (13).

The increasing scarcity and increased cost of hand labor challenged many farmers to reduce their dependence on hand hoeing. Substitution of the use of herbicides for hand labor substantially reduced the cost of production in several crops (13). Hand labor is still used to some extent to supplement other weed control practices, but it is used primarily to remove only those weeds that have survived herbicidal treatment. It is unlikely that hand hoeing will be a major component of future weed control programs.

CHEMICALS

Development of selective herbicides has spearheaded the advances in weed control technology during the last 30 years. Probably more than 30 individual herbicidal treatments have been registered for each major crop in the United States (14). The array of selective herbicides provides farmers with a wide choice of practices for control of the broad spectrum of weeds in continuously shifting weed populations. Herbicides will continue to be a key ingredient in most integrated weed management systems in the foreseeable future (14).

CURRENT INTEGRATED WEED MANAGEMENT SYSTEMS

In the United States weed control in major row crops is presently accomplished by the judicious integration of hand, mechanical, and chemical control procedures. In most row crops some direct weed control is accomplished through primary tillage and also by the incorporation of specific herbicides. Additional tillage during the growing season is used basically to control weeds between the crop rows. Selective herbicides are applied in a band directly over the crop drill or as broadcast. Hand hoeing is limited basically to those weeds that are not controlled by herbicides and to those weeds that are very competitive (Tables 2–5).

The total cost for full-season weed control in an individual crop varies considerably. Total weed control investment per hectare in peanut and cotton are similar with costs of $168.63 and $156.28, respectively, whereas costs per hectare in corn and soybean stand at $95.00 and $74.09, respectively. Equipment and labor costs are much higher for cotton than soybean, corn, or peanut. The costs of herbicide inputs for peanut are much higher than for the other three crops. These weed control expenditures may seem high, but when compared to the total or net worth of an individual crop and the potential loss that may occur if weeds are not controlled, then it would be apparent that investments are essential to the economical production of these crops. It has been estimated that without

TABLE 2.   Estimated Cost of Full-Season Weed Control in Solid Cotton Grown on Sandy Soil[a]

Operation	1980 Cost ($/ha)			
	Equipment	Labor	Materials	Total
Disk (6.4 m); broadcast and incorporate 0.56 kg trifluralin/ha	9.14	1.38	7.78	18.30
Plant; apply 0.56 kg fluometuron/ha (50-cm band on 100-cm row spacing)	0	0	6.50	6.50
Cultivate	7.68	1.61	0	9.29
Cultivate; postdirect 0.9 kg MSMA/ha (50-cm band on 100-cm row spacing)	10.00	1.83	2.79	14.62
Cultivate; postdirect 0.33 kg cyanazine + 1.12 kg MSMA/ha (50-cm band on 100-cm row spacing)	10.00	1.83	5.71	17.54
Hand hoe (4.9 hours/ha)	0	15.32	0	15.32
Cultivate; postdirect 1.12 kg of MSMA + 0.45 kg fluometuron/ha (50-cm band on 100-cm row spacing)	7.51	1.35	8.67	17.53
Cultivate; postdirect 1.12 kg of MSMA + 0.22 kg diuron/ha (50-cm band on 100-cm row spacing)	7.51	1.35	4.87	13.73
Hand hoe (4.9 hours/ha)	0	15.32	0	15.32
Cultivate; postdirect 1.12 kg linuron/ha	7.51	1.35	19.27	28.13
Total	59.35	41.34	55.59	156.28

Source: Constructed from data furnished in Ref. 15.

[a]Based on Mississippi Delta practices and six-row equipment in 1980.

TABLE 3. Estimated Cost of Full-Season Weed Control in Soybeans Grown on Clay Soil[a]

Operation	1980 Cost ($/ha)			
	Equipment	Labor	Materials	Total
Disk (6.4 m); broadcast and incorporate 1.12 kg trifluralin/ha	9.14	1.38	15.57	26.09
Plant; apply 0.45 kg metribuzin/ha (50-cm band on 100-cm row spacing)	0	0	7.39	7.39
Cultivate	7.68	1.61	0	9.29
Postdirect 0.11 kg 2,4-DB + 0.25 kg linuron/ha (50-cm band on 100-cm row spacing)	5.09	1.38	6.23	12.70
Cultivate	5.14	1.06	0	6.20
Cultivate; postdirect 0.67 kg dinoseb/ha (50-cm band on 100-cm row spacings)	7.51	1.38	3.53	12.42
Total	34.56	6.81	32.72	74.09

Source: Reference 15.
[a]Based on Mississippi Delta practices and six-row equipment in 1980.

the use of herbicides, in the United States the losses due to weeds in cotton, corn, peanut, and soybean would be 40, 25, 90, and 24%, respectively (18).

COST AND EXTENT OF HERBICIDE USAGE

Herbicides were used on 9.3 million ha of agricultural land in 1949, on 12.1 million ha in 1952, on 21.5 million ha in 1959, and on 28.7 million ha in 1962 (14). Herbicides were used on 48.6 million ha in 1965, on 81 million ha in 1975, and on about 101 million ha in 1977 (14). It is possible that in 1980 herbicides were used on more than 121 million ha.

In 1979 the herbicide sales represented 43% of the 1978 worldwide pesticide sales (19). In the United States herbicides represented 61% of all pesticide sales (Table 6).

In 1978 herbicide sales in corn (38%) and soybean (32%) accounted for 70% of all herbicide sales in the United States. Herbicide sales in all

TABLE 4. Estimated Cost of Full-Season Weed Control in Peanuts Grown on Sandy Soil[a]

Operation	1980 Cost ($/ha)			
	Equipment	Labor	Materials	Total
Disk (3.6 m); broadcast and in-corporate 1.68 kg benefin + 2.24 kg vernolate/ha[b]	9.98	2.57	51.25	63.80
Cracking stage; apply broadcast 3.36 kg alachlor + 1.68 kg dinoseb + 3.36 kg nap-talam/ha	6.35	3.16	51.05	60.56
Cultivate	8.62	2.96	0	11.58
Postbroadcast 0.84 kg/ha 2,4-DB[c]	6.35	3.16	6.86	16.37
Postbroadcast 0.84 kg/ha 2,4-DB	6.35	3.16	6.86	16.37
Total	37.65	15.01	116.02	168.68

Source: Ref. 16.

[a]Based on Georgia Practices and four-row equipment in 1980.

[b]On approximately 35% of the hectarage the herbicides are incorporated with a power-driven rotary tiller when the beds are formed. The equipment cost is $14.83/ha.

[c]4-(2,4-Dichlorophenoxy)butanoic acid.

TABLE 5. Estimated Cost of Full-Season Weed Control in Corn Grown on Silt Loam Soil[a]

Operation	1980 Cost ($/ha)			
	Equipment	Labor	Materials	Total
Disk (6.4 m); broadcast and in-corporate 4.48 kg butylate + 2.24 kg atrazine/ha	13.59	2.74	38.30	54.63
Field cultivator (8.5 m)	9.88	2.13	0	12.01
Rotary hoe (7.3 m)	4.94	0.54	0	5.48
Cultivate	11.12	3.50	0	14.62
Postbroadcast 0.56 kg/ha 2,4-D	2.47	1.98	3.81	8.26
Total	42.00	10.89	42.11	95.00

Source: Ref. 17.

[a]Based on Illinois practices and six-row equipment in 1980.

TABLE 6. Herbicide and Total Pesticide Purchases (in $ million) by Crop in the United States and the World in 1978

Crop	Total Herbicide Purchases		Total Pesticide Purchases	
	United States	World	United States	World
Corn	658	940	871	1286
Cotton	128	297	348	1222
Wheat	69	346	87	498
Sorghum	50	71	70	130
Rice	37	307	54	854
Other grain	21	184	30	264
Soybean	550	696	593	810
Tobacco	13	27	35	116
Peanut	26	41	70	125
Sugarbeet	19	194	32	290
Sugarcane	10	86	19	143
Other field crops	17	64	45	168
Alfalfa	13	22	40	74
Other hay and forage crops	7	12	11	32
Pasture and rangeland	28	64	33	78
Fruits and nuts	39	127	227	1190
Vegetables	46	140	166	701
Other crops	—	98	—	303
Total	1731	3716	2731	8284

Source: Ref. 19.

other crops were considerably less than those in corn and soybean: cotton, 7%; wheat, 4%; sorghum, 3%; rice, 2%; other grain crops, about 1%; peanut, 1%; sugarbeet, 1%; and pasture and rangeland, 1%. The total for all horticultural crops, fruits, and vegetables was 5%. Forty-seven percent of all herbicides sold in 1978 were in the United States (Table 6).

The value of all herbicides marketed worldwide in 1978 was estimated at about $3.7 billion (19). The sale of herbicides for weed control in corn in the United States represented 25% of the worldwide herbicide sales, whereas the United States' share of the herbicide sale in soybean accounted for 19% of the worldwide sales. Wheat, rice, cotton, and horticultural crops accounted for 9, 8, 8, and 7%, respectively, of the worldwide herbicide sales (Table 6).

Worldwide herbicide sales were estimated to increase by 13.4% from 1980 to 1984 (19). The projected increase in herbicide sales in the United

States from 1980 to 1984 was estimated at about 15%. It was estimated that by 1984 herbicide sales would represent 59% of all pesticide sales in the United States, whereas herbicides would account for 53% of the pesticide sales worldwide (19).

Nearly 200 herbicides are registered for use in the United States, but triazine (30.5%) and amide (29.6%) herbicides represent about 60% of all sales (20). The phenoxy herbicides accounted for about 11% of the herbicides used in 1976, and the carbamates accounted for about 10.2% of sales. All other chemical groups represented only a very small portion of the market in 1976, but the dinitroaniline herbicides have increased in usage at a rapid rate since that time.

In corn, cotton, rice, soybean, and peanut more than 80% of the 1976 hectarage in the United States was treated with at least one application of a herbicide (Table 7). The average quantity of herbicides applied annually on these five crops was 2.87 kg/ha, and an average across all crops was 2.24 kg/ha. A total of 178.99 million kg of herbicides was applied in 1976 in the United States, with 53% of this total applied to corn and 21% to soybean.

PROBLEMS ASSOCIATED WITH CONVENTIONAL WEED CONTROL

Significant progress has been made in the development of conventional weed control technology, but weeds continue to cause severe reductions in crop yield and quality. With changes in production technology, such as the use of cropping sequences, rotations, cultivations, chemical, and non-chemical methods, ecological shifts in the weed spectrum are likely. Even with the most effective weed control methods available, there are still serious weed problems that remain unsolved in the production of individual crops (7,21-27).

TROUBLESOME WEEDS

Johnsongrass is the worst weed in cotton and causes about 16% of the total cotton losses caused by weeds (Table 8). The perennial johnsongrass can be controlled postemergence only by hand applications of herbicides or by hand-hoeing. Johnsongrass is listed as the weed most difficult to control in more states than any other weed in cotton (28,29). Seedling johnsongrass and nutsedge in cotton are controlled with either disodium methanearsonate (DSMA) or MSMA applied postemergence, but these

TABLE 7.   Total Area of Major Field Crops, Hay and Forage Crops, and Pasture and Rangeland, Percentage of Hectares Treated with Herbicides, and Total and Per Hectare Herbicide Usage in the United States in 1976

Crop	Hectares Grown (Million)	Percentage of Hectares Treated with Herbicides	Herbicides Applied (Active Ingredients)	
			Total (Million kg)	Per Hectare (kg)
Corn	34.0	90	93.9	3.0
Cotton	4.7	84	8.3	2.1
Wheat	32.4	38	9.9	0.7
Sorghum	7.5	51	7.1	1.9
Rice	1.0	83	3.8	4.4
Other grain	12.0	35	2.4	0.5
Soybean	20.3	88	36.7	2.0
Tobacco	0.4	55	0.5	2.2
Peanut	0.6	93	1.5	2.6
Alfalfa	10.7	3	0.7	1.2[a]
Other hay and forage crops	13.9	2	—	—
Pasture and rangeland	197.5	1	4.3	3.0
Total/average	335.0	22	169.1[b]	2.6
Total average excluding pasture and rangeland	137.5	56	164.8	2.2

Source: Ref. 20.

[a] Alfalfa, other hay, and forage collectively.

[b] Includes 9.25 million kg of herbicides used on other crops.

herbicides do not kill the underground reproductive portions of the plants. Therefore, repeated applications of these herbicides are necessary.

In addition to johnsongrass, the worst weeds in cotton are common cocklebur, prickly sida, morningglories, and nutsedge (Table 8). These weeds cause nearly 60% of all cotton losses due to weeds (28,29). These weeds are difficult to control with either preplant or preemergence herbicide treatments. Common cocklebur, prickly sida, and morningglories are resistant to many of the postemergence herbicides used in cotton. Unfortunately, there are no selective herbicides that can be applied over-the-top for the control of these weeds.

Pigweed, fall panicum, croton, spurred anoda, and sicklepod are

TABLE 8. Percent of Cotton Yield Losses Caused by the Five Worst Weeds in Cotton Within a State from 1974 to 1976

Cropping Regions and States	Level of Yield Losses Caused by Weeds (%)				
	Johnson-grass	Common Cocklebur	Prickly Sida	Morning-glories	Nutsedge
Southeast					
Alabama	13	20	20	11	13
Georgia	—[a]	30	10	—	20
North Carolina	3	20	22	12	10
South Carolina	5	10	15	12	10
Average	5.3	20	16.8	8.8	13.3
Midsouth					
Arkansas	20	—	50	10	1
Louisiana	14	10	20	11	16
Missouri	—	70	1	20	—
Mississippi	25	—	35	2	10
Tennessee	40	14	16	6	6
Average	19.8	18.8	24.4	9.8	6.6
Southwest					
Oklahoma	18	7	—	5	8
Texas	25	5	—	10	1
Average	21.5	6.0	—	7.5	4.5
West					
Arizona	—	—	—	—	—
California	20	3	—	17	15
New Mexico	30	—	—	25	15
Average	25.0	1.5		21.0	15.0
United States— average	16.4	14.5	14.5	10.8	9.6

Source: Adapted from Ref. 28.

[a]A dash indicates that no information was available.

ranked 6–10 as worst weeds in cotton (28,29). These cause about 12% of the losses in cotton compared with losses of 66% by the five worst weeds mentioned previously. Pigweed is most troublesome in the Southeast, Southwest, and West. Fall panicum, croton, and sicklepod are troublesome only in the Southeast. Spurred anoda appears to be spreading throughout the Cotton Belt. Pigweed, fall panicum, croton, spurred anoda, and sicklepod are all difficult to control and will increase in severity unless better control techniques are developed.

The list of the worst weeds in soybean is even more diverse than that in cotton because soybean is produced over a larger geographic area (30). In the mid-Atlantic region, crabgrass, fall panicum, redroot pigweed, common cocklebur, and annual morningglories are the five most common weeds (Table 9). The five most costly weeds to control in the mid-Atlantic region are jimsonweed, yerba-de-tago, spurred anoda, giant foxtail, and sicklepod.

In the Southeast, common cocklebur, sicklepod, annual morningglories, pigweeds, and johnsongrass are most common; sicklepod, common cocklebur, annual morningglories, johnsongrass, and pigweeds are the most costly to control (30).

TABLE 9.   Five Most Common and Costly Weeds in Soybean

Most Common Weeds	Most Costly Weeds
Mid-Atlantic Region	
Crabgrass	Jimsonweed
Fall panicum	Yerba-de-tago
Redroot pigweed	Spurred anoda
Common cocklebur	Giant foxtail
Annual morningglories	Sicklepod
Southeastern Region	
Common cocklebur	Sicklepod
Sicklepod	Cocklebur
Annual morningglories	Annual morningglories
Pigweeds	Johnsongrass
Johnsongrass	Pigweeds
Delta Region	
Johnsongrass	Johnsongrass
Common cocklebur	Prickly sida
Annual morningglories	Common cocklebur
Barnyardgrass	Nutsedge
Sesbania	Annual morningglories
North Central Region	
Giant foxtail	Common cocklebur
Pigweeds	Velvetleaf
Smartweed	Milkweed
Velvetleaf	Yellow nutsedge
Common cocklebur	Giant foxtail

Source: Ref. 21.

In the Delta states, the five most common weeds are johnsongrass, common cocklebur, annual morningglories, barnyardgrass, and hemp sesbania. The most costly weeds to control are johnsongrass, prickly sida, common cocklebur, nutsedges, and morningglories (Table 9).

In the north central states, the five most common weeds are giant foxtail, pigweeds, smartweed, velvetleaf, and common cocklebur. The most costly weeds to control are cocklebur, velvetleaf, milkweeds, yellow nutsedge, and giant foxtail (30).

All major crops are plagued with a variety of weed problems such as those described previously and new weed control problems appear almost annually. It is especially difficult to control weeds in reduced tillage systems, and ecological shifts of weeds occur much more rapidly under reduced tillage than with conventional tillage. Weeds that are expected to cause much greater damage in the future under reduced tillage systems in soybeans include hemp dogbane, Jerusalem artichoke, climbing milkweed, hedge bindweed, field bindweed, common milkweed, wirestem muhly, horsenettle, and trumpetcreeper. In cotton, weeds that are increasing in severity include bermudagrass, spurges, horsenettles, groundcherry, velvetleaf, jimsonweed, ragweed, bristly starbur, Texas panicum, signalgrass, sunflower, goosegrass, buffalobur, barnyardgrass, and Texas blueweed. Many of these presently cause only small yield reductions but are listed as very difficult to control.

It has been estimated that weeds cause an average annual loss of about $8 billion in the United States (Table 10). This is about 12.4% of the

TABLE 10. Estimated Average Annual Monetary Losses Due to Weeds in the United States from 1973 to 1977

Commodity Group	Average Annual Monetary Losses (in $1000)
Field crops	5,735,821
Vegetables	450,093
Fruits and nuts	299,498
Forage seed crops	38,763
Hay	676,221
Pasture and rangelands	788,805
Total	7,989,201

Source: Ref. 31.

potential value of crops produced (31). The losses caused by weeds and the cost of their control in the United States are about $14 billion annually (5). About 72% of the weed losses occur in field crops and about 18%, in forage crops. About 10% of the losses due to weeds occur in vegetables, fruits, and nut crops (31). These losses will continue to increase unless new methods of weed control are developed.

HERBICIDE RESIDUES IN SOIL

The herbicides used in crop production are broken down biologically and chemically. Only a few, such as some of the substituted ureas and triazines, degrade slowly enough to be carried over into the next growing season (32). Nevertheless, care is needed to ensure that recommended rates are not exceeded. There is a continuing need to evaluate each new herbicide carefully so that future problems with herbicide residues are avoided.

POSSIBLE AIR AND WATER POLLUTION

The likelihood of drift of spray droplets and herbicide vapors will be increased without adequate education and supervision of the users. Some herbicides may be lost to the atmosphere during or after application. The volatility of many herbicides is low, and educational programs may further aid in reducing problems concerning volatility. Droplet drift of herbicides can be avoided by spraying under calm weather conditions or when light winds are blowing away from susceptible crops. Educational programs may also aid in preventing drift problems by emphasizing low spray pressures and the use of adjuvants to increase droplet size.

There is always the inherent possibility of water pollution when herbicides are applied. Herbicides may be inadvertently applied directly to water or reach water indirectly through runoff from the soil surface. Most herbicides are rapidly adsorbed on the surface of soil particles or organic matter, thus minimizing runoff. The occurrence of heavy rainfall before adsorption occurs may result in increased runoff. Fortunately, herbicides in runoff water have not posed a serious problem so far. Although herbicides are occasionally detected in water, they occur usually in very low concentrations (33).

There have been a few reports of herbicides in groundwater, and these infrequent reports have not shown groundwater pollution by leaching to be either extensive or significant.

CROP INJURY

Crops may be injured following the use of herbicides. Injury usually occurs when a herbicide is applied at an excessive rate or when an adverse environmental condition follows application. Crop injury problems occur every year, but these represent a very low percentage of the total herbicide applications.

Some crop injury problems relate to the rotation of both crops and herbicides. The herbicides used in corn and cotton, for example, may injure soybean planted during the following year. Injury of this type is most common when the growing season is very dry during the year in which the herbicides were applied. With some herbicides like the triazines, crop injury may result from insufficient breakdown of the herbicide in soil. With triazines, adsorption depends primarily on soil acidity. Therefore, the soil pH may influence herbicide persistence and the consequent phytotoxicity to the crop. Rates of herbicide breakdown in soil may also depend on the properties and formulation of the herbicide, microbial populations, soil moisture levels, type of soil management, extent and intensity of rainfall, and soil and air temperatures. All these factors may affect crop injury from herbicides. Therefore, the problems of herbicide injury to crops are often complex, but most crop injury problems relating to persistence can be avoided by close adherence to the instructions provided on the herbicide label.

HERBICIDE RESIDUES IN CROPS

The presence of significant residues of herbicides in harvested crops is quite rare. Usually this problem is studied and satisfactory answers are found during the course of generation of the vast amount of data required for registration of a new herbicide for sale. Also, monitoring and surveillance programs of food for sale provide continuing evidence that pesticide residues do not exceed acceptable levels. There is, however, a constant need to assure the public that there is no risk from either herbicide residues in food or from the impurities in herbicide formulations.

HUMAN EXPOSURE DURING HERBICIDE APPLICATIONS

Herbicides as a class of chemicals have a low degree of toxicity to mammals. Presently, few herbicides are in the restricted-use categories. However, herbicides are potentially dangerous to those who come in direct contact with them, and adequate precautions should always be taken to minimize exposure. Repeated educational programs are needed to teach users to avoid skin contact or inhalation of spray droplets, dusts, or vapors. Safe storage of herbicides is needed for prevention of accidental poisoning.

HERBICIDE COMPATIBILITY

The use of tank mixtures of two or more herbicides or mixtures of herbicides and other pesticides is increasing in frequency. Tank mixtures require registration prior to such use. The use of untested mixtures causes some concern because of the greater probability of crop injury, decrease in weed control, or formation of unstable mixtures that interfere with efficient application. More research is needed to ensure that herbicide mixtures are safe and reliable.

DISPOSAL OF HERBICIDE CONTAINERS

The increased use of herbicides has also increased the disposal problem for used containers. Used containers should not be reused, and proper disposal is troublesome. There is no immediate solution to this problem that will satisfy all concerned, and more research is needed to solve this problem.

FUTURE RESEARCH NEEDS IN WEED SCIENCE

The development of practical measures of biological weed control is the primary need for more effective integrated weed management systems. Comparatively little effort is being devoted to research aimed at developing biological weed control methods. Biological control techniques appear to be cost-effective and should aid in reducing total weed control costs. At present, biological control is possible for a few weeds, but future biological weed control methods could have far-reaching benefits.

There is a need for more research to determine the comparative toxicity of herbicides to different crop cultivars and to breed cultivars with increased resistance to herbicides. Recent research has shown differential susceptibility of cotton, corn, and soybean cultivars to herbicides. It seems likely that in the future herbicide-resistant cultivars will become more important as the hectarage treated with herbicides increases.

There has been little work to establish the exact extent of land area infested with individual weeds, the level of infestation, or the rate of spread of individual weeds. There has been too little research on competition of weed populations that justify the use of specific control measures. An understanding of weed population dynamics will enable better prediction of ecological changes, especially with regard to influences by chemical and nonchemical treatments. More information is also needed on methods to improve efficacy of herbicide treatment and to aid in prevention of undesirable shifts in weed populations that will make weed control more difficult.

Allelopathic compounds are those secondary chemicals released by plants that cause direct or indirect effects on other plants. Research on the role of secondary compounds on plant interactions is in its infancy. Allelopathic compounds will probably play a major role in development of new methods for controlling all weed pests (34). Future research on allelopathy will require close cooperation between chemists and biological scientists, and the potential for protection of crop plants from natural allelopathic toxicants appears to be great.

There is also a need for more research on factors that affect weed seed germination. Uniform germination of weed species through the use of germination stimulants will greatly enhance the chances for effective weed control. The availability and use of improved seed germination stimulants will lead to the development of weed control systems that could be used during intervals when the crops are not growing.

Research is needed to develop herbicides or formulations with controlled-release characteristics. Advances in controlled-release technology will help to reduce the rates of application needed, reduce volatility, reduce herbicide movement through soil profile, increase crop selectively, and reduce environmental exposure. The successful development of controlled-release technology could revolutionize the use of herbicides in agriculture.

Improvements are needed in the design and efficiency of herbicide application equipment. Newer ways to apply herbicides more economically with greater safety to crops and to environment are urgently needed.

The differential biological and ecological requirements of weeds and crops are poorly understood. There are only a few examples where these requirements have been adequately exploited for weed control. Differences in life histories of plants have not been fully investigated and have been used only incidentally for improved control. Ecological shifts in weed populations as affected by repeated use of herbicides occurs more frequently in crop rotations, and the cause for these is poorly understood.

Research has shown that crops can be protected against herbicides by treating crop seeds with protective chemicals. The basis for this protection is largely unknown. Exploitation of protection through treatment of crop seeds could result in the use of previously nonselective herbicides in highly cost-effective control programs.

The use of reduced tillage in crop production has increased in recent years, but the reduced tillage practice is often restricted by the lack of better weed control programs. Minimum and zero tillage practices conserve fuel, water, and soil; reduce herbicide volatility; and reduce losses of herbicide and sediment sheet erosion in runoff. There is a need to determine which herbicide treatments are most effective in reduced tillage systems.

CONCLUSIONS

Weeds continue to cause annual losses of about 12% in American agricultural production, creating a total loss of $14 billion annually. The most successful weed control programs are those that combine the use of crop competition and rotation and chemical, mechanical, and other control techniques in an integrated weed management system. Many exciting new control techniques are available for future progress, but many of these, including the biological control measures, have been explored to only a limited extent.

The most important development in weed control during the past two decades is the discovery and development of safe and effective new herbicides that selectively control specific weeds in major crops.

There are many opportunities for practical breakthroughs by improving our basic understanding of weed ecology, life cycles of weeds, and weed seed germination. New weed controls may also result from studies on allelopathic compounds, genetic characteristics of weeds and crops, and new herbicide application techniques. However, massive shifts from the use of herbicides to nonchemical technology are not likely in the foreseeable future. The new technology developed on nonchemical control will likely be most effective when integrated with conventional technology.

REFERENCES

1. O. C. Burnside. 1979. Weeds. Pp. 27–38 in: W. B. Ennis, Jr., ed., *Introduction to Crop Protection*. American Society of Agronomy, Madison, WI.
2. L. G. Holm, D. L. Plucknett, J. V. Pancho, and J. P. Herberger. 1977. *The World's Worst Weeds: Distribution and Biology*. The University Press of Hawaii, Honolulu, HI, 609 pp.
3. Anonymous. 1965. *Losses in Agriculture*. Agriculture Handbook No. 291. U.S. Department of Agriculture. U.S. Government Printing Office, Washington, DC, 120 pp.
4. Anonymous. 1972. *Extent and Cost of Weed Control with Herbicides and an Evalua-*

tion of Important Weeds, 1968. U.S. Department of Agriculture, ARS-H-1. U.S. Government Printing Office, Washington, DC, 227 pp.

5. Anonymous. 1976. *Weed Control Technology for Protecting Crops, Grazing Lands, Aquatic Sites, and Noncropland.* ARS-NRP No. 20280. U.S. Department of Agriculture. U.S. Government Printing Office, Washington, DC, 185 pp.

6. C. G. McWhorter. 1978. *Disciplinary Challenges: Weed Science.* Special Report No. 67. University of Arkansas, Fayetteville, AR, 114 pp.

7. C. G. McWhorter and D. T. Patterson. 1980. Ecological factors affecting weed competition in soybeans. Pp. 371–392 in: F. T. Corbin, ed., *World Soybean Research Conference II: Proceedings.* Westview Press, Boulder, CO, 897 pp.

8. C. G. McWhorter. 1976. P. 14 in: *Report of the ARS Research Planning Conference on Weed Control in Soybeans.* Interdepartmental Communication, U.S. Department of Agriculture, Science and Education Administration, Southern Weed Science Laboratory, Stoneville, MS, 54 pp.

9. C. G. McWhorter, 1977. Pp. 12–14 in: *Reports of the ARS Research Planning Conference on Weed Control in Cotton.* Interdepartmental Communication, U.S. Department of Agriculture, Science and Education Administration, Southern Weed Science Laboratory, Stoneville, MS, 36 pp.

10. Anonymous. 1980. *Weed Control Guidelines for Mississippi.* Mississippi Agricultural and Forestry Experiment Station, Mississippi State, MS, 152 pp.

11. J. M. Chandler and L. R. Oliver. 1979. *Spurred Anoda: A Potential Weed in Southern Crops.* U.S. Department of Agriculture, Science and Education Administration. Agriculture Review and Manual, Southern Series, No. 2, 19 pp.

12. J. E. Dale and J. M. Chandler. 1979. Herbicide-crop rotation for johnsongrass (*Sorghum halepense*) control. *Weed Sci.* **27**: 479–485.

13. A. F. Wiese and J. M. Chandler. 1979. Weeds. Pp. 232–238 in: W. B. Ennis, Jr., ed., *Introduction to Crop Protection.* American Society of Agronomy, Madison, WI, 524 pp.

14. W. C. Shaw. 1978. Herbicides: The cost/benefit ratio—the public view. *Proc. South. Weed Sci. Soc.* **31**: 28–47.

15. D. W. Parvin, Jr., J. G. Hamill, and F. T. Cooke, Jr. 1980. *Budget for Major Crops, Delta of Mississippi.* AECMR No. 95. Mississippi Agricultural and Forestry Experiment Station, Mississippi State, MS, 61 pp.

16. E. W. Hauser and G. Westberry. 1980. USDA-SEA-AR, Georgia Coastal Plain Experiment Station, Tifton, GA, personal communication.

17. L. M. Wax and R. H. Hinton. 1980. USDA-SEA-AR, University of Illinois, Urbana, IL, personal communication.

18. J. R. Abernathy. 1979. Nationwide crop losses caused by weeds without herbicide availability. *Abstr. Weed Sci. Soc. Am.*, p. 66.

19. Anonymous. 1979. A look at world pesticide markets. *Farm Chem.* **142** (9), 61–68.

20. T. R. Eichers, P. A. Andrilenas, and T. W. Anderson. 1978. *Farmers' Use of Pesticides in 1976.* Agricultural Economics Report 418. U.S. Department of Agriculture. U.S. Government Printing Office, Washington, DC, 58 pp.

21. C. C. Dowler and M. B. Parker. 1975. Soybean weed control systems in two southern costal plain soils. *Weed Sci.* **23**, 198–202.

22. E. W. Hauser, M. D. Jellum, C. C. Dowler, and W. H. Marchant. 1972. Systems of weed control for soybeans in the costal plain. *Weed Sci.* **20**, 592–598.

23. W. C. Shaw. 1979. Integrated weed management systems technology. Pp. 149–157 in: J. M. Brown, ed., *Proceedings of the Beltwide Cotton Production Research Conferences.* National Cotton Council, Memphis, TN.

24. F. W. Slife. 1979. Weed control systems in the corn belt states. Pp. 393–398 in: F. T. Corbin, ed., *World Soybean Research Conference II: Proceedings.* Westview Press, Boulder, CO, 897 pp.

25. F. W. Slife and L. M. Wax. 1976. Weed and herbicide management. Pp. 397–403 in: L. D. Hill, ed., *World Soybean Research: Proceedings.* The Interstate Printers and Publishers, Danville, IL, 1073 pp.

26. L. M. Wax. 1976. Difficult-to-control annual weeds. Pp. 420–425 in: L. D. Hill, ed., *World Soybean Research: Proceedings.* The Interstate Printers and Publishers, Danville, IL, 1073 pp.

27. L. M. Wax. 1977. New weed control developments. Pp. 41–44 in: *Proceedings of the Soybean Summit I.* American Soybean Association, St. Louis, MO, 72 pp.

28. D. V. DeBord. 1977. *Cotton Insect and Weed Loss Analysis.* The Cotton Foundation, Memphis, TN, 122 pp.

29. C. G. McWhorter and T. N. Jordan. 1982. Limited tillage in cotton production. In: *Reduced Tillage in Crop Production.* A monograph by the Weed Science Society of America, Champaign, IL, in press.

30. W. C. Shaw, L. M. Wax, C. G. McWhorter, C. C. Dowler, and R. N. Andersen. 1979. Progress in the development of weed control technology for soybean production. Pp. 50–63 in: R. Judd, ed., *Fifty Years with Soybeans.* National Soybean Crop Improvement Council. Hilton Head, SC.

31. J. M. Chandler. 1980. Assessing losses caused by weeds. Pp. 234–240 in: *Proceedings of the E. C. Stakman Commemorative Symposium.* Minnesota Agricultural Experiment Station Miscellaneous Publication No. 7.

32. J. M. Chandler and K. E. Savage. 1980. Phytotoxic interaction between phenylurea herbicides in a cotton (*Gossypium hirsutum*)-soybean (*Glycine max*) sequence. *Weed Sci.* **28:** 521–526.

33. J. H. Caro. 1976. Pesticides in agriculture runoff. Pp. 91–119 in: B. A. Stewart, ed., *Control of Water Pollution from Cropland.* Vol. II—*An Overview.* U.S. Department of Agriculture, ARS-H-5-2. U.S. Government Printing Office, Washington, DC, 187 pp.

34. Anonymous. 1977. *Report of the Research Planning Conference on the Role of Secondary Compounds in Plant Interactions (Allelopathy).* Mississippi State, MS, 124 pp.

3

STATUS OF WEED CONTROL WITH PLANT PATHOGENS

G. E. TEMPLETON
Department of Plant Pathology, University of Arkansas,
Fayetteville, Arkansas

Chapters 2 and 4 illustrate the need for biological weed control technology as well as the existence of natural regulation of plant populations by pathogens. What are the prospects for controlling some critical weed problems with plant pathogens? Certainly there is great diversity of opinion on this question among plant pathologists and weed scientists. Only time and massive efforts will provide the answer. As we conclude the initial decade of modest research and development of plant pathogens for weed control, it seems appropriate to examine the current status of the concept, consolidate the empirical evidence we have gained, reveal the perceived potential of various pathogens being tested to meet diverse weed control needs, and review the demonstrated potential of a few that have progressed to field implementation. This chapter discusses *where we are now, where we are going,* and *how we may best get there!*

STATUS OF RESEARCH

The task of examining efforts to control weeds with plant pathogens is simplified by reviews by Freeman et al. (1), Trujillo and Templeton (2), and Templeton and Smith (3) and a compilation of biocontrol projects in plant pathology by Charudattan (4). An updated composite of projects from these sources is listed in Table 1.

Although Table 1 may not be an exhaustive list, it serves to examine

TABLE 1. Status of Projects Studying the Potential for Biological Control of Aquatic and Terrestrial Weeds with Plant Pathogens

Weed	Pathogen	Location	Status[a]	Researcher(s)	Reference(s)
	Bacteria				
1. Algae (blue-green)	*Bdellovibrio bacteriovorus* Stolp and Starr	Ohio	C	Burnham	5
2. *Avena fatua* L. (wild oat)	*Pseudomonas syringae* van Hall	Montana	A	Sands	4
3. *Solanum dulcamara* L. (bitter nightshade)	*Pseudomonas solanacearum* E. F. Sm.	Montana	C	Sands	4
4. *Sorghum halepense* (L.) Pers. (johnsongrass)	*Pseudomonas syringae* van Hall	Arizona	C	Stangellini	6
	Fungi				
1. *Acroptilon repens* (L.) DC. (= *Centaurea repens* L.) (Russian knapweed)	*Puccinia acroptili* Syd.	Russia	G	Kovalev et al.	7
2. *Aeschynomene virginica* (L.) B. S. P. (northern jointvetch)	*Colletotrichum gloeosporioides* (Penz.) Sacc. f. sp. *aeschynomene*	Arkansas	E	Daniel et al. Templeton et al.	8 9
3. *Ageratina riparia* (Regel) King and Robinson (= *Eupatorium riparium* Regel) (hamakua pamakani)	*Cercosporella ageratinae* (nomen nudem)	Hawaii	F	Trujillo	10
4. *Albizzia julibrissin* Durazzini (silktree albizzia)	*Fusarium oxysporum* Schlecht. f. sp. *perniciosum* (Hept.) Toole	Hawaii	E	Gardner	4
5. *Alternanthera philoxeroides* (Mart.) Griseb. (alligatorweed)	*Alternaria alternantherae* Holcomb and Antonopoulos	Louisiana	C	Holcomb and Antonopoulos	11
6. *Ambrosia artemisiifolia* L. (common ragweed)	*Puccinia xanthii* Schw. *Albugo tragopogonis* Pers.	Russia Canada	C C	Kovalev Watson	1 12
7. *Anoda cristata* (L.) Schlecht. (spurred anoda)	*Alternaria macrospora* Zimm.	Mississippi	D	Walker and Sciumbato	13
8. *Arceuthobium* spp. (dwarf mistletoes)	*Wallrothiella arceuthobii* (Peck) Sacc.	Oregon	G	Knutson and Hutchins	14
	Nectria fuckeliana Booth var. *macrospora*	Canada	D	Funk et al.	15
	Colletotrichum gloeosporioides von Arx	California	G	Parmeter et al.	16
	Various Fungi	Colorado	D	Hawksworth et al.	17
9. *Avena fatua* L. (wild oat)	*Puccinia coronata* Corda	Montana	C	Sands	18
10. *Brasenia schreberi* Gmel. (watershield)	*Dichotomophthoropsis nymphaearum* (Rand) M. B. Ellis	Minnesota	C	Johnson and King	19

30

TABLE 1. (*Continued*)

Weed	Pathogen	Location	Status[a]	Researcher(s)	Reference(s)
	Fungi				
11. *Cannabis sativa* L. (hemp)	*Fusarium oxysporum* Schlecht. f. sp. *cannabis* Snyder and Hans.	California	G	McCain	20
12. *Cassia surrattensis* Lamarck (brushweed)	*Cephalosporium* sp.	Hawaii	D	Trujillo and Obrero	21
13. *Centaurea solstitialis* L. (yellow starthistle)	*Puccinia jaceae* Otth	Maryland	A	Emge	4
14. *Chondrilla juncea* L. (skeletonweed)	*Puccinia chondrillina* Bubak and Syd.	Australia	F	Cullen Hasan	22 23
		California	E	Emge and Kingsolver Weinhold Dunkle	24 4 4
	Erysiphe cichoracearum DC. ex Merat	France (CSIRO, Australia)	C	Hasan	25
	Leveillula taurica (Lév.) Arnaud	France (CSIRO, Australia)	C	Hasan	25
15. *Cirsium arvense* (L.) Scop. (Canada thistle)	*Fusarium roseum* (Link) Snyder and Hans.	Wyoming	C	Rai and Bridgemon	26
	Puccinia obtegens (Link) Tul.	Montana	D	Faye and Sands	27
	Alternaria spp.	Montana	D	Faye and Sands	27
16. *Convolvulus arvensis* L. (field bindweed)	Various	Canada	B	Harris	1
17. *Cuscuta campestris* Yunck. *Cuscuta cupulata* Engelm.	*Alternaria cuscutacidae* Rudakov	Russia	D	Miusov and Bashaeva	28
Cuscuta campestris Yunck. *Cuscuta epithymum* Murr. (dodders)	*Colletotrichum destructivum* O'Gara	Oregon	G	Leach	29
18. *Cyperus esculentus* L. (yellow nutsedge)	*Puccinia canaliculata* (Schw.) Lagh.	California	C	McCain	4
	Sclerotinia homoeocarpa Bennett	Indiana	D	Brown	4
Cyperus rotundus L. (purple nutsedge)	*Puccinia* sp.	Philippines	G	Ou	4
	Phyllosticta sp.		G	Ou	4
19. *Diospyros virginiana* L. (persimmon)	*Cephalosporium diospyri* Crandell	Oklahoma	E	Griffith	30
20. *Eichhornia crassipes* (Mart.) Solms (water-hyacinth)	*Acremonium zonatum* (Saw.) Gams	Florida	D	Freeman et al. Charudattan et al.	1,31 32
	Alternaria eichhorniae Nag Raj and Ponnappa	India	G	Nag Raj and Ponnappa	33

31

TABLE 1.   (*Continued*)

Weed	Pathogen	Location	Status[a]	Researcher(s)	Reference(s)
	Fungi				
	Bipolaris stenospila Drechs.	Florida	G	Charudattan et al.	34
	Cercospora piaropi Tharp.	Florida	C	Freeman and Charudattan	35
	Cercospora rodmanii Conway	Florida	E	Conway et al.	36–40
	Fusarium roseum (Link) Sacc.	Florida	G	Freeman et al.	1
	Myrothecium roridum Tode ex Fr.	Indonesia	C	Setyawati	1
	Rhizoctonia sp.	Florida	D	Freeman	41
				Freeman and Zettler	42
		Indonesia	C	Setyawati	1
	Uredo eichhorniae Gonz.-Frag. and Cif.	Florida	C	Charudattan and Conway	43
21. *Eupatorium adenophorum* Spr. (Crofton weed or pamakani)	*Cercospora eupatorii* Peck	Australia	F	Dodd	44
22. *Hydrilla verticillata* (L. fil.) Royle (hydrilla)	*Sclerotium rolfsii* (Sacc.) Curzi	Florida	G	Charudattan	45
	Pythium sp.	Florida	G	Charudattan	45
	Penicillium sp.	Florida	G	Charudattan and Lin	46
	Aspergillus sp.	Florida	G	Charudattan and Lin	46
	Trichoderma sp.	Florida	G	Charudattan and Lin	46
	Phytophthora parasitica Dast.	Florida	G	Freeman et al.	1
	Fusarium solani (Mart.) Appel and Wollenw.	Indiana	D	Brown	4
	Fusarium roseum 'Culmorum' (Link ex. Fr.) Snyder and Hans.	Florida	D	Charudattan et al.	47,48
23. *Hydrocotyle umbellata* L. (water pennywort)	*Cercospora hydrocotyles* Ellis and Everh.	Florida	G	Conway	4
24. *Ipomoea hederacea* (L.) Jacq. (ivyleaf morningglory)	*Coleosporium ipomoeae* (Schw.) Burr.	Arkansas	C	TeBeest	49
25. *Jussiaea decurrens* (Walt.) DC. (winged waterprimrose)	*Colletotrichum gloeosporioides* (Penz.) Sacc. f. sp. *jussiaeae*	Arkansas	E	Boyette et al.	50
26. *Moluccella* sp. (= *Molucella*) (molluccabalm)	*Cercospora molucellae* Bremer and Petrak	Israel	C	Kenneth	4

TABLE 1. (*Continued*)

Weed	Pathogen	Location	Status[a]	Researcher(s)	Reference(s)
	Fungi				
27. *Morrenia odorata* Lindl. (strangler vine or milkweed vine)	*Aecidium asclepiadinum* Speg.	Florida	C	Charudattan et al.	51
	Puccinia araujae Lév.	Florida	C	Charudattan et al.	51
	Phytophthora citrophthora (R. E. Sm. and E. H. Sm.) Leonian	Florida	E	Burnett et al. Ridings et al.	52 53
28. *Nuphar luteum* (L.) Sm. (yellow water-lily)	*Cercospora nymphaeacea* Cooke and Ellis	Florida	G	Conway	4
29. *Nymphaea odorata* Ait. (fragrant waterlily)	*Dichotomophothoropsis nymphaearum* (Rand) M. B. Ellis	Minnesota	C	Johnson and King	19
	Nymphaea tuberosa Paine (white water-lily) *Dichotomophothoropsis nymphaearum* (Rand) M. B. Ellis	Minnesota	C	Johnson and King	19
30. *Nymphoides orbiculata* (Sm.) Ktze. (water-lily)	Various fungi	Sweden	C	Nilsson	4
31. *Orobanche* spp. (broomrapes)	*Fusarium oxysporum* Schlecht. var. *orthoceras* (Appel and Wollenw.) Bilay	Russia	F	Kovalev	54
32. *Oxalis* sp. (woodsorrel)	*Puccinia oxalidis* (Lév.) Diet. and Ellis	France	C	Durrieu	55
33. *Panicum dichotomiflorum* Michx. (fall panicum)	*Sorosporium cenchri* Henn.	Pennsylvania	D	Zeiders	4
34. *Populus alba* L. (aspen or white poplar)	*Valsa* spp.	Minnesota	D	French	4
35. *Quercus* spp. (red- and bur oaks)	*Ceratocystis fagacearum* (Bretz) Hunt	Minnesota	D	French and Schroeder	56
36. *Rubus constrictus* Lef. and M. *Rubus ulmifolius* Schott. (wild blackberries)	*Phragmidium violaceum* (Schultz) Wint.	Chile	E	Oehrens and Gonzales	57
37. *Rumex crispus* L. (curly dock)	*Uromyces rumicis* (Schum.) Wint.	Italy	G	Inman	58
38. *Salvinia* spp. (salvinias)	*Alternaria* sp. *Spicariopsis* sp.	Rhodesia	G	Loveless	59
39. *Sida spinosa* L. (prickly sida)	*Colletotrichum malvarum* (A. Braun and Casp.) Southworth	Arkansas	C	TeBeest	49
40. *Sorghum halepense* (L.) Pers. (johnsongrass)	*Sphacelotheca cruenta* (Kuehn) Potter	California	C	Lindow	60
	Helminthosporium spp.	Mississippi	C	Walker	61

TABLE 1. *(Continued)*

Weed	Pathogen	Location	Status[a]	Re-searcher(s)	Ref-eren-ce(s)
Fungi					
41. *Xanthium canadense* Mill. (cocklebur)	*Puccinia canaliculata* (Schw.) Lagh.	California	C	McCain	4
42. *Xanthium pungens* Wallr. (Noogoora burr)	*Puccinia xanthii* Schw.	Australia	C	Alcorn Hasan	62 23
Xanthium strumarium L. (heartleaf cockle-bur)	*Puccinia xanthii* Schw.	North Carolina	D	Van Dyke	63
Nematodes					
1. *Acroptilon repens* (L.) DC. (= *Centaurea repens* L.) (Russian knapweed)	*Paranguina picridis* Kirjanova and Ivanova	Canada	D	Watson	64
2. *Myriophyllum spicatum* L. (eurasian water-milfoil)	*Aphelenchoides fragariae* (Ritz.-Boz) Christie	Florida	G	Smart and Esser	65
3. *Solanum elaeagnifolium* Cav. (silverleaf night-shade)	*Nothanguina phyllobia* (Thorne) Thorne	Texas	D	Orr et al.	66
Viruses					
1. Algae (blue-green)	Viruses (cyanophages)	Various	C	Jackson and Sladecek Cannon	67 68
2. *Chondrilla juncea* L. (skeletonweed)	Virus (?)	France (CSIRO, Australia)	C	Hasan et al.	69
3. *Morrenia odorata* Lindl. (strangler vine or milkweed vine)	Araujia mosaic virus	Florida	C	Charudattan et al.	70
4. *Myriophyllum spicatum* L. (eurasian water-milfoil)	Virus (?)	Maryland	G	Bayley	71
5. *Pistia stratiotes* L. (waterlettuce)	Virus	Nigeria	G	Pettet and Pettet	72
6. *Solanum carolinense* L. (horsenettle)	Tobacco mosaic virus	Russia	F	Izhevsky	73

[a]Status of projects determined from the researchers' most recent report. A—planning stage; B—surveys have been undertaken; C—studies in laboratory/greenhouse in progress; D—under small-scale field tests; E—under large-scale field tests; F—operational (the biological control agent is in deliberate use or is exerting a natural control); and G—inactive project or status unknown.

what individual researchers have perceived to be good target-biocontrol agent systems for exploration, evaluation, development, and eventual field use. Projects are listed from 13 countries, spanning climatic zones from the temperate to the tropic. In the United States, 18 states are represented (see recapitulation of geographic area in Table 2).

Weed species listed in Table 1 include algae, submerged aquatics, floating aquatics, semiaquatics, terrestrial annuals and perennials, herbaceous and woody species, crop mimics, poisonous plants, and parasitic plants. Infestations listed cover waterways, pastures and rangelands, annual cultivated crops, orchard crops, roadways, and forests. Many of these weeds are generally considered intractable problems because of their ability to spread rapidly over large areas and produce dense infestations on land or waterways of economic value too low to warrant conventional weed control techniques. On the other hand, several are weeds in intensively cultivated, high-value crops that lack adequate chemical controls or that need more environmentally responsible techniques than are now available.

Pathogens listed for these 54 weed species number approximately 83 and include fungi, bacteria, nematodes, and viruses. They range from facultative parasites to obligate parasites. Many are indigenous fungal pathogens, but several introduced pathogens are also being studied intensively. Generally, introduced pathogens have been considered for biocontrol of widespread, introduced weeds in areas of low economic value.

TABLE 2. Geographic Distribution of Projects on Biological Control of Weeds with Plant Pathogens

Non-U.S. Countries		
Australia	Indonesia	Philippines
Canada	Israel	Rhodesia
Chile	Italy	Russia
France	Nigeria	Sweden

U.S. States		
Arizona	Louisiana	Ohio
Arkansas	Maryland	Oklahoma
California	Minnesota	Oregon
Florida	Mississippi	Texas
Hawaii	Montana	W. Virginia
Indiana	North Carolina	Wyoming

A summary of pathogen groups examined is presented in Table 3. The number of fungi considered far exceeds that of other pathogen groups. This is not surprising since most pathogens of higher plants are fungi. Other biological attributes and technological advantages of fungi that contribute to their preeminence in biological weed control are discussed in other chapters.

The classes of fungal pathogens studied or considered for weed controls are listed in Table 4. The number of deuteromycetes or *fungi imperfecti* studied substantially exceeds that of other classes. Again, this is not surprising since this group is among the more commonly encountered classes of fungi found as pathogens on higher plants, and the members of this group can be easily cultured and induced to sporulate.

It can be concluded from the foregoing analysis of projects under way or being considered that we are still in an exploratory phase of the science. Greatest emphasis is being placed on the more apparent candidate pathogens in a fortuitous or opportunistic approach. A more systematic approach would surely improve our prospects (74) and appears warranted

TABLE 3.   Pathogen Classes Being
Studied for Biological Weed Control

Pathogen Groups	Number of Species or Strains
Fungi	71 +
Bacteria	3
Nematodes	3
Viruses	6
Total	83 +

TABLE 4.   Classes of Fungal
Pathogens Studied for Weed Control

Class of Fungi	Number of Projects
Myxomycetes	0
Phycomycetes	4
Ascomycetes	7
Deuteromycetes	38
Basidiomycetes	18
Total	67

from the empirical evidence gained in a few of the more advanced projects (75,76).

BIOCONTROL TACTICS

Two distinct tactics apparently are emerging: the classical or traditional tactic and the bioherbicide tactic (3,75). The classical tactic is directed principally at plants that have been introduced into a new region or country and become weedy in the absence of their natural enemies. Pathogens are sought from the geographic origins of the plants for introduction into new regions with the expectation that they will become established, increase to epiphytotic levels, and eventually become endemic when the weed is suppressed to subeconomic levels. A variant of this rationale is the importation of pathogen strains from the same or related host species for perpetuation on indigenous weed species that have evolved in the absence of the pathogen and thus have little or no genetic resistance to it. This rationale derives from experience with such historically significant diseases as chestnut blight, white pine blister rust, coffee rust, and potato late blight, which were caused by accidentally introduced fungal pathogens that wrought drastic changes in cultivated or native plant populations.

The bioherbicide tactic, on the other hand, is directed at indigenous plants that have become weeds for one reason or another. Some may have reached an equilibrium state with their natural enemies that is above economically acceptable levels. Others may have become weedy as a result of human activities such as cultivation, fertilization, including enrichment of waterways with nutrient pollution; selective pesticide usage; or some crop, range, or pasture management practice that disrupts or eliminates the pathogen life cycle. Pathogens are sought among the indigenous weeds, particularly in undisturbed areas, that may be grown and induced to sporulate in large-scale fermentation tanks. They are applied, usually annually, as herbicides so that each plant is inundated with inoculum at the most propitious time for infection. Infection and pathogenesis thus create an epidemic with a pathogen that normally persists at an endemic level. Epidemics of these diseases may occur naturally in undisturbed areas, because environments (including host density) favorable to development occur only periodically, or conversely, because the environment may be usually favorable but the pathogen is irregular in its occurrence as a result of poor overwintering capacity, weak saprophytic ability, lack of vectors, or other constraints to dissemination. An understanding of these principles of plant pathology and weed biology has been employed in some of the more advanced projects.

SOME NOTABLE EXAMPLES

Skeletonweed has been successfully controlled in Australia by introduction of the rust fungus *Puccinia chondrillina* from the Mediterranean region where the weed originated. Estimated annual savings at final equilibrium between the fungus and skeletonweed equal $25.96 million (22). Attempts to duplicate this achievement in the western United States are being followed with great interest. Pamakani weed (*Ageratina riparia*) has been successfully controlled in Hawaii by introduction of the deuteromycete *Cercosporella ageratina* from the Caribbean area, the origin of the weed (10). Intense public sector research is needed to seek and evaluate pathogens for classical biological control of some of our more serious introduced weeds such as leafy spurge, Canada thistle, hemp sesbania, waterhyacinth, alligatorweed, and johnsongrass.

The bioherbicide, or more specifically, the mycoherbicide approach with indigenous fungal pathogens has proved biologically feasible and is approaching commercial use in at least three cases. Waterhyacinth control by application of *Cercospora rodmanii* is being pilot-tested over a large area by the U.S. Army Corps of Engineers, and the fungus is expected to be marketed by Abbott Laboratories in the near future. Likewise, Abbott Laboratories, in cooperation with the Florida Department of Agriculture, has recently registered the fungus *Phytophthora palmivora* (previously thought to be *P. citrophthora*) as a mycoherbicide for control of strangler vine in Florida citrus groves. Commercial development of an indigenous fungus, *Colletotrichum gloeosporioides* f. sp. *aeschynomene*, for control of northern jointvetch in rice and soybean is being field-tested in Arkansas in a cooperative effort among the U.S. Department of Agriculture, the state, and industry. The fungus is expected to be available from The Upjohn Company pending approval by the Environmental Protection Agency in 1982, more than a decade after its discovery.

We must be encouraged by the prospects for additions to our weed control arsenal from the pathogens listed in Table 1. A realistic appraisal of the potential of plant pathogens as weed controls in future pest management systems is not easy at this point. We simply do not have sufficient number of examples as to the feasibility of various pathogen-weed combinations, nor do we have enough information about those projects that have been considered but as yet have not reached the point of practical use. Are the projects active? If they are inactive, why? Did they suffer from lack of support, or did biological limitations, such as genetic variability in pathogen or host, slow growth of the pathogen, restricted sporulation, low virulence, or too broad a host range of pathogen lead to their abandonment? Was there conflict of interest in the control of the target species? Some projects may have failed as a result of organizational difficulties such as the lack of cooperation between plant pathologists and weed scientists or the absence of proper interfacing with industry at a critical phase in the development of the project. Simple economic constraints in product development coupled with limited market potential of highly specific agents may have discouraged development by industry. We need to know more about projects that have been or are being considered but have not proceeded to practical use in order to guide research, exploration, and selection of the most promising pathogens from the many pathogens available on most weed species.

CONCLUSIONS

In summary, the current status of biological weed control with plant pathogens may be aptly characterized as a fluid, progressive one with increased interest among plant pathologists, weed scientists, public agencies, fermentation scientists, and industries. The needs for additional inputs into weed control technology are apparent to all. A glimpse of the potential of plant pathogens in weed control can be envisioned from the empirical successes of a few. Lack of success in some pathogen-weed combinations, although not clearly documented, is reasonable to expect from either biological constraints or economic deterrents. Both success and lack of success reaffirm the uniqueness with which we are dealing in each disease and host biology combination. It is this uniqueness that presents the challenge of discovery and innovation to researchers in pathology, weed science, and fermentation. There are several challenges ahead: to find effective pathogens among those previously described or new; to develop combinations of pathogens that either more effectively kill individual weeds or serve to broaden the spectrum of weeds that may be killed by one application; to use aerial pathogens as damping off incitants as being evaluated by Walker (77); to enhance virulence in the pathogen directly or by chemical modification of the host physiology; to exploit the capacity of certain pathogens to suppress growth of plants or reduce seed formation of weeds; and to combine pathogens and insects for weed control.

It is our belief that cautious optimism among scientists in this area is justified and is well founded on sound biological principles. We may expect increasing interest and continued progress toward practical utilization of plant pathogens in future weed control technology.

REFERENCES

1. T. E. Freeman, R. Charudattan, and K. E. Conway. 1978. Status of the use of plant pathogens in the biological control of weeds. Pp. 201–206 in: T. E. Freeman, ed., *Proceedings of the Fourth International Symposium on Biological Control of Weeds*. University of Florida, Gainesville, FL.

2. E. E. Trujillo and G. E. Templeton. 1981. The use of plant pathogens in biological control of weeds. Pp. 345–350 in: D. Pimentel, ed., *Agricultural Handbook Series: Integrated Pest Management*. Chemical Rubber Company Press, Boca Raton, FL.

3. G. E. Templeton and R. J. Smith, Jr. 1977. Managing weeds with pathogens. Pp. 167–176 in: J. G. Horsfall and E. B. Cowling, eds., *Plant Disease: An Advanced Treatise*. Vol. 1. *How Disease Is Managed*. Academic Press, New York.

4. R. Charudattan. 1978. *Biological Control Projects in Plant Pathology—A Directory*. Miscellaneous Publication, Plant Pathology Department, University of Florida, Gainesville, FL, 67 pp.

5. J. C. Burnham. 1975. Bacterial control of aquatic algae. Pp. 120–125 in: D. L. Brezonik and J. L. Fox, eds., *Proceedings of Symposium on Water Quality Management Through Biological Control*. Department of Environmental Engineering Sciences, University of Florida, Gainesville, FL.

6. M. E. Stangellini. 1980. Department of Plant Pathology, University of Arizona, Tucson, AZ, personal communication.

7. O. V. Kovalev, L. G. Danilov, and T. S. Ivanova. 1973. Method of controlling Russian knapweed (in Russian). *Opisanie Izorbriteniia Kavtorskomu Svidetel 'stuv Byull*. 38, 2 (translation). Translation Bureau, Canada Department, Secretary of State, Ottawa, Canada.

8. J. T. Daniel, G. E. Templeton, R. J. Smith, Jr., and W. T. Fox. 1973. Biological control of northern jointvetch in rice with an endemic fungal disease. *Weed Sci.* **21**, 303–307.

9. G. E. Templeton, D. O. TeBeest, and R. J. Smith, Jr. 1978. Development of an endemic fungal pathogen as a mycoherbicide for biocontrol of northern jointvetch in rice. Pp. 214–216 in: T. E. Freeman, ed., *Proceedings of the Fourth International Symposium on Biological Control of Weeds*. University of Florida, Gainesville, FL.

10. E. E. Trujillo. 1980. Department of Plant Pathology, University of Hawaii, Honolulu, HI, personal communication.

11. G. E. Holcomb and A. A. Antonopoulos. 1976. *Alternaria alternantherae:* A new species found on alligatorweed. *Mycologia* **68**, 1125–1129.

12. A. K. Watson. 1980. Plant Science Department, McGill University, Sainte Anne DeBellevue, P. Q., Canada, personal communication.

13. H. L. Walker and G. L. Sciumbato. 1979. Evaluation of *Alternaria macrospora* as a potential biocontrol agent for spurred anoda (*Anoda cristata*): Host range studies. *Weed Sci.*, **27**, 612–614.

14. D. M. Knutson and A. S. Hutchins. 1979. *Wallrothiella arceuthobii* infecting *Arceuthobium douglasii:* Culture and field inoculation. *Mycologia* **71**, 821–828.

15. A. Funk, R. B. Smith, and J. A. Baranyay. 1973. Canker of dwarf mistletoe swellings on western hemlock caused by *Nectria fuckeliana* var. *macrospora. Can. J. For. Res.* **3**, 71–74.

16. J. R. Parmeter, Jr., J. R. Hood, and R. F. Scharpf. 1959. Colletotrichum blight of dwarf mistletoe. *Phytopathology* **49**, 812–815.

17. F. G. Hawksworth, E. F. Wicker, and R. F. Scharpf. 1977. *Fungal Parasites of Dwarf Mistletoes*. U.S. Department of Agriculture, Forest Service. General Technical Report RM-36, 14 pp.

18. D. C. Sands. 1980. Department of Plant Pathology, Montana State University, Bozeman, MT, personal communication.

19. D. A. Johnson, and T. H. King. 1976. A leaf-spot disease of three genera of aquatic plants in Minnesota. *Plant Dis. Rep.* **60**, 726–730.

20. A. H. McCain. 1978. The feasibility of using Fusarium wilt to control marijuana. *Phytopathol. News* **12**(9), 129.

21. E. E. Trujillo and F. P. Obrero. 1978. Cephalosporium wilt of *Cassia surattensis* in Hawaii. Pp. 217–220 in: T. E. Freeman, ed., *Proceedings of the Fourth International Symposium on Biological Control of Weeds*. University of Florida, Gainesville, FL.

22. J. M. Cullen. 1978. Evaluating the success of the programme for the biological control of *Chondrilla juncea* L. Pp. 117–121 in: T. E. Freeman, ed., *Proceedings of the Fourth International Symposium on Biological Control of Weeds*. University of Florida, Gainesville, FL.

23. S. Hasan. 1974. Recent advances in the use of plant pathogens as biocontrol agents of weeds. *PANS* **20**, 437–443.

24. R. G. Emge and C. H. Kingsolver. 1977. Biological control of rush skeleton weed with *Puccinia chondrillina*. *Proc. Am. Phytopathol. Soc.* **4**, 215.

25. S. Hasan. 1972. Behavior of the powdery mildews of *Chondrilla juncea* L. in the Mediterranean. *Actas Terceivo Congr. da Uniao Fitopatol. Mediterr. Oeiras* **1972**, 171.

26. I. S. Rai and G. H. Bridgmon. 1971. Studies on root rot of Canada thistle (*Cirsium arvense*) caused by *Fusarium roseum*. *J. Colorado-Wyoming Acad. Sci.* **17**, 2.

27. P. Faye and D. C. Sands. 1980. Department of Plant Pathology, Montana State University, Bozeman, MT, personal communication.

28. I. N. Miusov and E. G. Bashaeva. 1968. Mass production of the fungus *Alternaria* sp., a parasite of field dodder (in Russian). *Vest. sel'.-khoz. Nauki, Alma-Ata* **11**, 87–90.

29. C. M. Leach. 1946. A disease of dodder caused by the fungus *Colletotrichum destructivum*. *Plant Dis. Rep.*, **42**, 827–829.

30. C. A. Griffith. 1980. Noble Foundation, Agricultural Division, Rt. 1, Ardmore, OK, personal communication.

31. T. E. Freeman, R. Charudattan, K. E. Conway, and F. W. Zettler. 1976. *Biological Control of Aquatic Weeds with Plant Pathogens*. Contract Report A-76-2, U.S. Army Corps of Engineers, Waterways Experiment Station, Vicksburg, MS, 21 pp.

32. R. Charudattan, B. D. Perkins, and R. C. Littell. 1978. Effects of fungi and bacteria on the decline of anthropod-damaged waterhyacinth (*Eichhornia crassipes*) in Florida. *Weed Sci.* **26**, 101–107.

33. T. R. Nag Raj and K. M. Ponnappa. 1970. Blight of waterhyacinth caused by *Alternaria eichhorniae* sp. nov. *Trans. Br. Mycol. Soc.* **55**, 123–130.

34. R. Charudattan, K. E. Conway, and T. E. Freeman. 1976. A blight of waterhyacinth, *Eichhornia crassipes* caused by *Bipolaris stenospila* (*Helminthosporium stenospilum*). *Proc. Am. Phytopathol. Soc.* **2**, 65.

35. T. E. Freeman and R. Charudattan. 1974. Occurrence of *Cercospora piaropi* on waterhyacinth in Florida. *Plant Dis. Rep.* **58**, 277–278.

36. K. E. Conway. 1976. *Cercospora rodmanii*, a new pathogen of waterhyacinth with biological control potential. *Can. J. Bot.* **54**, 1079–1083.

37. K. E. Conway. 1976. Evaluation of *Cercospora rodmanii* as a biological control of waterhyacinths. *Phytopathology* **66**, 914–917.

38. K. E. Conway and T. E. Freeman. 1977. Host specificity of *Cercospora rodmanii*, a potential biological control of waterhyacinth. *Plant Dis. Rep.* **61**, 262–266.

39. K. E. Conway, T. E. Freeman, and R. Charudattan. 1978. Development of *Cercospora rodmanii* as a biological control for *Eichhornia crassipes*. Pp. 225–230 in: Anonymous, ed., *Proceedings of the European Weed Research Society Fifth Symposium on Aquatic Weeds*. P. O. Box 14, Wageningen, The Netherlands.

40. K. E. Conway, T. E. Freeman, and R. Charudattan. 1978. Method and compositions for controlling waterhyacinth. U.S. Patent 4,097,261.

41. T. E. Freeman. 1975. Rhizoctoniosis of aquatic plants. Pp. 327–328 in: *McGraw-Hill Yearbook of Science and Technology*, McGraw-Hill, New York.

42. T. E. Freeman and F. W. Zettler. 1971. Rhizoctonia blight of waterhyacinth. *Phytopathology*, **61**, 892.

43. R. Charudattan and K. E. Conway. 1975. Comparison of *Uredo eichhorniae*, the waterhyacinth rust, and *Uromyces pontederiae*. *Mycologia* **67**, 653–657.

44. A. P. Dodd. 1961. Biological control of *Eupatorium adenophorum* in Queensland. *Aust. J. Sci.* **23**, 356–365.

45. R. Charudattan. 1973. Pathogenicity of fungi and bacteria from India to hydrilla and waterhyacinth. *Hyacinth Control J.* **11**, 44–48.

46. R. Charudattan and C. Y. Lin. 1974. Isolates of *Penicillium*, *Aspergillus* and *Trichoderma* toxic to aquatic plants. *Hyacinth Control J.* **12**, 70–73.

47. R. Charudattan and D. E. McKinney. 1978. A Dutch isolate of *Fusarium roseum* 'Culmorum' may control *Hydrilla verticillata* in Florida. Pp. 219–224 in: Anonymous, ed., *Proceedings of the European Weed Research Society Fifth Symposium on Aquatic Weeds*. P. O. Box 14, Wageningen, The Netherlands.

48. R. Charudattan, T. E. Freeman, R. E. Cullen, and F. M. Hofmeister. 1980. Evaluation of *Fusarium roseum* 'Culmorum' as a biological control for *Hydrilla verticillata*: Safety. Pp. 307–323 in: E. S. Del Fosse, ed., *Proceedings of the Fifth International Symposium on Biological Control of Weeds*. CSIRO, Canberra, Australia.

49. D. O. TeBeest. 1980. Department of Plant Pathology, University of Arkansas, Fayetteville, AR, personal communication.

50. C. D. Boyette, G. E. Templeton, and R. J. Smith, Jr. 1979. Control of winged water-primrose (*Jussiaea decurrens*) and northern jointvetch (*Aeschynomene virginica*) with fungal pathogens. *Weed Sci.* **27**, 497–501.

51. R. Charudattan, H. A. Cordo, A. Silveira-Guido, and F. W. Zettler. 1978. Obligate pathogens of the milkweed vine, *Morrenia odorata* as biological control agents. P. 241 in: T. E. Freeman, ed., *Proceedings of the Fourth International Symposium on Biological Control of Weeds*. University of Florida, Gainesville, FL.

52. H. C. Burnett, D. P. H. Tucker, and W. H. Ridings. 1974. Phytophthora root and stem rot of milkweed vine. *Plant Dis. Rep.* **58**, 355–357.

53. W. H. Ridings, C. L. Schoulties, N. E. El-Gholl, and D. J. Mitchell. 1977. The milkweed vine pathotype of *Phytophthora citrophthora* as a biological control agent of *Morrenia odorata*. *Proc. Internatl. Soc. Citriculture* **3**, 877–881.

54. O. V. Kovalev. 1977. Biological control of weeds (in Russian). *Zashch. Past.* **22**, 12–14.

55. G. Durrieu. 1978. *Puccinia oxalidis*, a help in the control of *Oxalis*. P. 241 in: T. E.

Freeman, ed., *Proceedings of the Fourth International Symposium on Biological Control of Weeds*. University of Florida, Gainesville, FL.

56. D. W. French and D. B. Schroeder. 1969. The oak wilt fungus, *Ceratocystis fagacearum* as a selective silvicide. *For. Sci.* **15**, 198–203.

57. E. Oehrens and S. Gonzales. 1974. Introduction of *Phragmidium violaceum* (Schulz) Winter as a biological control agent for zarzamora (*Rubus constrictus* Lef. and M. and *R. ulmifolius* Schott.) (in Spanish). *Agro Sur (Chile)* **2**, 30–33.

58. R. E. Inman. 1971. A preliminary evaluation of *Rumex* rust as a biological control agent for curly dock. *Phytopathology* **61**, 102–107.

59. A. R. Loveless. 1969. The possible role of pathogenic fungi in local degeneration of *Salvinia auriculata* Aublet on Lake Kariba. *Ann. Appl. Biol.* **63**, 61–69.

60. S. E. Lindow. 1980. Department of Plant Pathology, University of California, Berkeley, CA, personal communication.

61. H. L. Walker. 1980. Spurred anoda [*Anoda cristata* (L.) Schlecht.] biocontrol with a plant pathogen. *Proc. South. Weed Sci. Soc.* **33**, 65.

62. J. L. Alcorn. 1975. Rust on *Xanthium pungens* in Australia. *Aust. Plant Pathol. Soc. Newsl.* **4**, 14–15.

63. C. G. Van Dyke. 1980. Department of Botany, North Carolina State University, Raleigh, NC, personal communication.

64. A. K. Watson. 1976. The biological control of Russian knapweed with a nematode. Pp. 221–223 in: T. E. Freeman, ed., *Proceedings of the Fourth International Symposium on Biological Control of Weeds*. University of Florida, Gainesville, FL.

65. G. C. Smart and R. P. Esser. 1968. *Aphelenchoides fragariae* in aquatic plants. *Plant Dis. Rep.* **52**, 455.

66. C. C. Orr, J. R. Abernathy, and E. B. Hudspeth. 1975. *Nothanguina phyllobia*, a nematode parasite of silverleaf nightshade. *Plant Dis. Rep.* **59**, 416–418.

67. D. F. Jackson and V. Sladecek. 1970. Algal viruses—eutrophication control potential. *Yale Sci. Mag.* **44**, 16–22.

68. R. Cannon. 1975. Field and ecological studies on blue-green algal viruses. Pp. 112–117 in: P. L. Brezonik and J. L. Fox, eds., *Proceedings of Symposium on Water Quality Management Through Biological Control*. Department of Environmental Engineering Sciences, University of Florida, Gainesville, FL.

69. S. Hasan, J. Giannotti, and C. Vago. 1973. Viruslike particles associated with a disease of *Chondrilla juncea*. *Phytopathology* **63**, 791–793.

70. R. Charudattan, F. W. Zettler, H. A. Cordo, and R. G. Christie. 1980. Partial characterization of a potyvirus infecting the milkweed vine, *Morrenia odorata*. *Phytopathology* **70**, 909–913.

71. S. E. M. Bayley. 1970. The Ecology and Disease of Eurasian Watermilfoil (*Myriophyllum spicatum* L.) in the Chesapeake Bay. Ph.D. thesis. Johns Hopkins University, Baltimore, MD, 190 pp.

72. A. Pettet and S. J. Pettet. 1970. Biological control of *Pistia stratiotes* L. in Western State, Nigeria. *Nature* **226**, 282.

73. S. S. Izhevsky. 1981. The application of pathogenic microorganisms for control of weeds in the USSR. Pp. 35–36 in: J. R. Coulson, ed., *Proceedings of the Joint American–Soviet Conference on the Use of Beneficial Organisms in the Control of Crop Pests*. Entomological Society of America, 4603 Calvert Road, College Park, MD.

74.   G. E. Templeton, R. J. Smith, Jr., and W. Klomparens. 1980. Commercialization of fungi and bacteria for biological control. *Biocontrol News Inf.* 1, 291–294.

75.   G. E. Templeton, D. O. TeBeest, and R. J. Smith, Jr. 1979. Biological weed control with mycoherbicides. *Annu. Rev. Phytopathol.* 17, 301–310.

76.   R. G. Emge and G. E. Templeton. 1981. Biological control of weeds with plant pathogens. Pp. 219–226 in: G. C. Papavizas, ed., *Proceedings of Beltsville Symposia in Agricultural Research 5. Biological Control in Crop Production.* Allanheld, Osmun and Company, Totowa, NJ.

77.   H. L. Walker. 1981. Granular formulation of *Alternaria macrospora* for control of spurred anoda (*Anoda cristata*). *Weed Sci.* 29, 342–345.

II

HOST-PARASITE
INTERACTION

4

IMPACT OF DISEASES ON PLANT POPULATIONS

P. C. QUIMBY, JR.

Southern Weed Science Laboratory, U.S. Department of Agriculture, Science and Education Administration, Agricultural Research, Stoneville, Mississippi

The objective of this chapter is to provide information regarding potential benefits and precautions that are related to the use of plant diseases for weed control. Specifically, this chapter is intended (1) to provide information about some historically significant epiphytotics and focus attention on the awesome potential of plant pathogens to alter plant populations, (2) to glean some insight from history to help explain why plant pathology is directed primarily toward the reduction of plant diseases, (3) to provide information to support the need for caution in research aimed toward biocontrol of weeds with plant pathogens, and (4) to attempt to relate the epiphytotic events to factors that apparently favored their occurrence.

As Ohr (1) mentioned, authors of most ecology reference books have largely ignored plant diseases as important determinants of the constituency of plant communities. One exception was Harper (2), who included a chapter on pathogens but wrote, "Pathogens have rarely been studied as part of an ecological system in nature, and their relevance to population biology and population genetics is mainly unexplored territory." Another recent exception is a review by Burdon and Shattock (3), who considered the effects of pathogens on the population dynamics of plants in monoculture and in stands of mixed species. They also examined the effect of plants on the population dynamics of pathogens. However, considerable information exists for specific epiphytotics, and this information is highly relevant to the use of pathogens for the biological control of weeds (see also Chapter 8).

ESSENTIAL REQUIREMENTS FOR EPIPHYTOTICS

The *Random House Dictionary* (4) defines "epiphytotic" as "the widespread destructive outbreak of a plant disease." Rangaswami (5) lists the following factors as essential to the occurrence of an epiphytotic (epidemic): "(a) presence of a large population of susceptible hosts in a vulnerable condition, (b) presence of a large population of the pathogen which is aggressive and virulent, (c) effective mode of quick dispersal of the pathogen to long distances, (d) ability of the pathogen to establish quickly and reproduce in large numbers, (e) not too exacting food requirements of the pathogen to grow and reproduce, and (f) favorable environmental conditions for infection and spread of the disease."

A different view was emphasized by van der Plank (6), namely, that an epidemic does not require an aggressive pathogen that reproduces rapidly, spreads over great distances quickly, and is not too strict in terms of requirements. An example provided is viral swollen shoot disease of cacao (6), which shows that an important epiphytotic can occur merely through plant contact. Thus the density of plants in a population can be an important factor. This relationship emphasizes the vulnerability of monocultures (see also Chapter 8).

SELECTED EPIPHYTOTICS

POTATO LATE BLIGHT IN IRELAND

The cultivated potato originated in the Andes mountains of South America. However, no one knows for certain when or where the late blight pathogen *Phytophthora infestans* originated because it is now present in most of the potato growing areas of the world. The likely home of the pathogen is Mexico (5), where wild, native *Solanum* spp. are hosts. Interesting accounts of this epiphytotic, which changed the history of nations, have been provided by Horsfall and Cowling (7) and Carefoot and Sprott (8).

Phytophthora infestans is a pythiaceous fungus that attacks foliage and tubers (5). The disease is favored by cool, wet, rainy weather.

The late blight invaded the potato fields of the United States in 1844 but caused little concern in Europe. In the 1840s the Irish potato fields were especially vulnerable. The Irish people depended almost solely on the potato for their food because the English landlords were exporting all of the grain. But in that terrible year of 1845, when temperatures dropped as much as seven degrees below normal and the cold rains began, many people lost their sole source of food to *P. infestans* within a month.

The weather conditions and potato blight played a repeat performance in 1846 and the increased hunger intensified three of humans' most dreaded diseases at that time—dysentery, typhus, and relapsing fever. Many who were still healthy fled Ireland, and the exodus continued for several years. Within 10 years 2 million or 25% of the Irish population had emigrated and a million people had died. Many of the emigrants came to America, where some encountered much ethnic and religious prejudice. However, the frontier of America needed their skills and strong backs to help build the railroads and western industry. Through the years, the descendants of these emigrants have been incorporated into the fabric of the citizenry so that now in the United States about 1 person in 10 has Irish genes.

In Europe the effects of the potato blight were far-reaching as the disease spread through Ireland and every other country on the continent to Russia. Although famine was not widespread, hunger and unrest were everywhere. In 1848 monarchies fell and republics were instituted. Louis Philippe of France abdicated; drastic constitutional changes were implemented in Austria and Germany. From this ferment emerged the influence of radicals such as Engels and Marx.

From our perspective of hindsight, it is difficult to realize that the cause of the potato blight was not even known at the time. The superstitious believed that the "little people" started it or that the "devil" had brought it. Another espoused the idea that the blight was caused by the steam locomotives that were rumbling about at 20 miles/hour and discharging electricity into the air. Others indicated that vapors from volcanoes were poisoning the potatoes.

In 1845 the Reverend Dr. Miles J. Berkeley observed the causal agent with his microscope and published the idea that this "mold" was causing the blight, but he offered no positive proof. Others regarded the mold as a saprophyte. In 1861 Anton deBary, now considered the father of modern plant pathology (5), proved that the fungus was the causal agent by exposing one group of plants to spores, thereby inducing the disease, while maintaining spore-free plants in a healthy condition (8). DeBary named the fungus *Phytophthora infestans*. By 1890 Bordeaux mixture (copper sulfate and lime) sprayed on potatoes reduced losses from the disease. A number of control methods are available today, but this disease is still important in potato growing areas of the world. The development of disease resistant potatoes is continuing today.

COFFEE RUST IN CEYLON

Hemileia vastatrix, a basidiomycete, is the causal agent of coffee rust. The history of this disease has been reported by Horsfall and Cowling (7) and Carefoot and Sprott (8). The disease apparently originated in Africa, the native home of coffee, and has had a profound effect on human culture. It turned Britain from a nation of coffee drinkers to a nation of tea drinkers.

Coffee was apparently brought by the Arabs from Ethiopia about 1000 A.D. Although the Arabs guarded the coffee beans, some were stolen, taken to India, and then carried from there to Ceylon (now Sri Lanka). Coffee was well adapted to Ceylon; in 1835 there were 203 ha of coffee which increased to 203,000 ha by 1870. Britons were coffee drinkers at that time. The annual coffee yield dropped from 50 million kg in 1870, when coffee rust first appeared, to virtually zero in 1889.

In 1889 Henry Marshall Ward was dispatched by the English government to Ceylon to develop a cure for the coffee rust. Although Ward distinguished himself in tracing the life history of the pathogen, his attempts to control the disease with sulfur were futile. This was before the discovery of the Bordeaux mixture. Bankrupt and desperate, the planters turned to tea and in 5 years had converted 136,000 ha from coffee to tea with 5500 to 9900 hand-planted tea bushes per hectare. This was a massive effort and a big gamble, since the market for tea was uncertain. Britain's loyalty to the planters resulted in a nation of tea drinkers. Today Britons consume nine cups of tea to one cup of coffee, despite the fact that a cup of tea contains only about one-third as much caffeine as a cup of coffee.

Subsequently, coffee cultivation has centered largely in South America (Brazil and Colombia), where until recently it has been free of the rust. In 1978 Horsfall and Cowling (7) reported the disquieting news that the rust is rapidly spreading in these countries. The ultimate effects are not known. Will Americans soon be joining their British counterparts in a tea break instead of a coffee break? History indicates that this is a possibility.

CHESTNUT BLIGHT IN THE UNITED STATES

An ascomycete, *Endothia parasitica*, the causal fungus of chestnut blight, is believed to have originated in China (2). Horsfall and Cowling (7) and

Kuhlman (9) have provided much information about this epiphytotic. The chestnut (*Castanea dentata*) was the dominant tree in large areas of the North American forest, but with the exception of isolated young or suckering plants, this species has been almost completely eliminated by the disease. "This is probably the largest single change in any natural plant population that has ever been recorded by man" (2).

In 1904 chestnut blight was discovered in the Bronx parks by H. W. Merkel, chief forester and founder of the New York Zoological Society. In 1905 Merkel trimmed the obvious disease out of 438 trees, but to no avail. Ninety-eight percent of all the chestnuts in the Bronx parks were infected in 1905. The disease, vectored by insects and birds, was extremely virulent and could girdle a smooth-barked 4-inch stem in 21 days. Rough barked limbs were more resistant; girdling occurred in 1–10 years.

The chestnut was a very valuable tree. It made up 25% of the eastern hardwood forest and ranged naturally over 91 million ha of land. The mature trees were magnificent, standing up to 37 m tall with boles of up to 2.1 m in diameter, and provided food and shelter for a diversity of wildlife. The wood was used for everything from roof shingles to fencing. The U.S. Forest Service estimated a value of $20 million for chestnut timber cut in 1909. In West Virginia alone standing chestnut was estimated at 10 billion board feet worth $25 million. Considerable expense and extensive research directed toward the control of chestnut blight were ineffective. One billion chestnut trees have disappeared. By 1915 all government programs were discontinued. The total cost of losses and attempted control was over $1 billion (8).

Hartline (10) has reported recently on biological control of chestnut blight with a viruslike disease of the blight-causing fungus. A hypovirulent strain of the fungus, weakened by the viruslike disease, is used to infest chestnut trees that then protects the trees from virulent strains. This approach has been more successful in Europe than in the United States. The American chestnut is very vulnerable to the virulent form of the fungus, which has diversified in the United States, and the hypovirulence approach seems to provide little or no control in North America (10).

DUTCH ELM DISEASE IN THE UNITED STATES

According to Carefoot and Sprott (8), Dutch elm disease, caused by the ascomycete *Ceratocystis ulmi*, was first discovered in the United States (Cleveland and Cincinnati) in 1930. One discoverer, Christine Buisman, a plant pathologist from Holland, reported that the disease had destroyed the elm trees in the Netherlands, and she named the causal agent *Ceratostomella ulmi*.

The mode of movement of the pathogen into Europe is unknown. Horsfall and Cowling (7) are proponents of the "wicker basket" theory; they suggest that the disease moved in the Chinese elm wicker baskets imported into Flanders with Chinese laborers during World War I. An exhaustive search of the literature provided no references to "wicker baskets," and movement of Chinese laborers into specific sites in Europe in 1916 and 1917 would not account for the tremendous spread of the disease in that time frame (11).

In America the importation of the disease was attributed (8) to furniture manufacturers who brought burled elm logs into the United States from Europe for veneer. The spread of the disease was greatly facilitated by the presence of two major vectors, the European elm bark beetle and the native elm bark beetle. The European elm bark beetle was introduced before the pathogen but was little noticed until it and the native beetle teamed up with the deadly fungus.

Dutch elm disease is a vascular wilt. The pathogen invades the xylem almost exclusively and causes dysfunction of the water-transporting tissue (12). The insects may "inoculate" the xylem directly (12), or the spores may enter the wounds caused by the beetles chewing into the bark to reach the phloem sap (8). The fungus penetrates the sapwood, and billions of spores that enter the xylem produce a toxic shock (12), which results in plugging and reduced conductivity in the xylem. The disease also spreads by root grafting from diseased trees to adjacent healthy trees (13).

In 1967 Carefoot and Sprott (8) estimated that half of the United States' 0.5 billion ornamental elm trees were already dead and predicted the ultimate demise of all the 1.5 billion elms in North America. They estimated the total cost of the losses and attempted control to be $50 billion.

JARRAH DIEBACK IN AUSTRALIA

Podger et al. (14) were the first to determine that a pythiaceous fungus, *Phytophthora cinnamomi*, is apparently the causal agent of jarrah dieback, a disease of *Eucalyptus marginata* in western Australia. The disease was first recorded in 1928, but the causal factor was verified only in 1965 by Podger et al. (14). This disease affects several plant species of the Cycadaceae, Xanthorrhoeaceae, Casuarinaceae, Proteaceae, and Myrtaceae in the jarrah forest. Destruction of the forest community has occurred on affected sites and the replacement vegetation might include less desirable species of *Eucalyptus* and ground flora consisting of species belonging to Restionaceae and Juncaceae.

Phytophthora cinnamomi can infect more than 190 species of plants and causes many plant diseases worldwide (15). This soilborne pathogen causes littleleaf disease of shortleaf and loblolly pine in the southeastern United States (16). In 1954 the disease caused mortality sufficient to interfere with forest management on 2.3 million ha or about 35% of the shortleaf pine growing area.

Avocado root rot is another disease caused by this pathogen (15). In 1967 avocado root rot had affected 2300 ha of avocado in California. As with jarrah dieback and littleleaf disease of pine, avocado root rot is associated with high rainfall or irrigation on poorly drained soils.

Native avocado trees of Latin America are resistant to the disease, but their fruit is too small for commercial production. Biocontrol of the disease has been obtained with alfalfa meal (17); the alfalfa meal apparently stimulates the buildup of a microbial antagonist that is effective against the disease-causing fungus.

BOTRYOSPHAERIA TAMARICIS ON SALTCEDAR IN ARGENTINA

Saltcedars, *Tamarix* spp., are phreatophytic Mediterranean plants introduced into the United States in the early 1800s (18). Watts et al. (18) have examined the potential for biocontrol of saltcedar in the southwestern United States, where the plant had invaded riparian habitats and displaced native (willow and cottonwood) phreatophytes.

Quimby et al. (19) reported that numberous herbicides have been tested for control of this aggressive brushy species. Hollingsworth et al. (20) evaluated several control techniques in field tests. The herbicidal methods developed are expensive and only a small percentage of the saltcedar is currently controlled chemically. An evaluation of biological control agents in the United States failed to uncover any natural enemies that provided significant control (18). Fungi that were isolated and identified appeared to be nonpathogenic, weak secondary invaders, or saprophytes (18).

For those of us who have observed the apparent health and vigor of saltcedar in the United States, the epiphytotic reported on saltcedar in Argentina by Frezzi (21) is of great interest. Apparently a 1.5- to 2-year-old planting of *Tamarix* was destroyed by the fungus *Botryosphaeria tamaricis* in Argentina. This may have been an isolated case as saltcedar still abounds in that country. Nevertheless, possibilities may exist for manipulating this fungus for biocontrol of saltcedar in Argentina.

OAK WILT IN THE UNITED STATES

Ceratocystis fagacearum, the causal agent of oak wilt, was originally described from Wisconsin and named *Chalara guercina* by Henry in 1944 (22). True et al. (22) have provided a very useful monograph on the subject in which they outline the available information on the origin of the pathogen. Apparently the fungus is not foreign as all exotic oaks tested were susceptible. Interestingly, the Chinese chestnut, imported to replace partially the decimated native chestnut, is extremely susceptible. Native white oaks are moderately resistant, thus suggesting that the wilt pathogen is native to North America.

Various hypotheses have been proposed to describe the oak wilt epiphytotic. One concept is that the fungus mutated from a preexisting saprophytic or weakly parasitic form or that the mutants spread more readily. Another suggestion is that one or more new vectors increased the speed and efficiency for dissemination of the pathogen. A third possibility is that the pathogen has always been present at its present level but was not recognized until recently. True et al. (22) did not subscribe to this last possibility, and they did not speculate as to the most likely theory. They pointed out that the highest disease incidence occurs in the mountainous regions of the Appalachians, where the nearly continuous forest land is covered by stands made up chiefly of oak species. Anderson and Skilling (23) estimated in 1955 that if oak wilt were to continue to spread in Wisconsin at its current rate, 21 to 30% of the oak woods will be invaded by the year 2055.

Lewis and Oliviera (24) have provided strong evidence that *C. fagacearum* is the primary cause of "live oak decline" in Texas. The fungus was consistently isolated from trees with wilt symptoms, and these isolates always produced disease symptoms when inoculated to healthy trees. According to Lewis and Oliviera (24), this disease was first reported on 200 trees near Austin, Texas in 1934 and had spread to 20 central Texas counties by 1949; however, the cause was unknown at that time.

Comparison of a 1957 map (22) of known oak wilt distribution with a 1980 map (25), reveals that the disease has spread southward into South Carolina; southwestward into southwestern Tennessee, southern Arkansas, Oklahoma, Kansas; and as far southwest as Kerrville, Texas (Fig. 1).

Tainter and Gubler (26) have reported that the range of oak wilt remained static in Arkansas from 1951 until 1973. They initially attributed the limited southward spread of oak wilt to poor growth of the fungus during high summer temperatures, but they acknowledged that *C. fagacearum* has the potential to spread far south of its present general occurrence as evidenced by the findings in Texas. Tainter and Gubler (26) have provided evidence that antagonistic fungi, namely, *Hypoxylon* spp. inhibit the spread of the disease.

IMPORTANCE OF COEVOLUTION IN THE OCCURRENCE OF EPIPHYTOTICS

IMPORTANCE OF COEVOLUTION TO EPIPHYTOTICS

Pianka (27) has provided a broad definition of coevolution: "coevolution refers to the joint evolution of two (or more) taxa that have close ecological relationships but do not exchange genes, and in which reciprocal selective pressures operate to make the evolution of either taxon partially dependent upon the evolution of the other."

In biogeographic sense, coevolution occurs as the interactions of native organisms tending toward homeostasis or internal stability within an ecological system. In several cases of the epiphytotics discussed in this chapter it is easy to see how the many hosts and pathogens were related, directly or indirectly, to extreme perturbations induced by humans. In many instances humans had transported pathogens to new sites and in other instances had introduced the host plants to new sites where they were exposed to different pathogens or where the pathogens were favored following the institution of monocultures.

For example, with potato late blight both the host and the pathogen were moved to new sites in Ireland where the cropping practices and weather favored the disease. In the example of the coffee rust, the pathogen and coffee plant were moved from Africa to Ceylon, where monoculture was practiced. The introduction of exotic pathogens decimated the North American chestnut and elm. Littleleaf of pine and avocado root rot have been serious diseases in North America. The origin of the causal organism *Phytophthora cinnamomi* is not known for certain, but the tendency toward monoculture of pine and the planting of Central American avocado in orchards of California have favored the progression of the disease. The same pathogen is considered an exotic in Australia, where it is responsible for jarrah dieback of *Eucalyptus* and other native Australian plant species.

Saltcedar, a native of the Mediterranean area and so refractory to all measures of control in the United States, was eliminated from a planting in Argentina by *Botryosphaeria tamaricis*. One can only speculate as to whether saltcedar encountered a new pathogen in Argentina or carried one of its own into a new environment.

The oak wilt pathogen is apparently native to North America, and

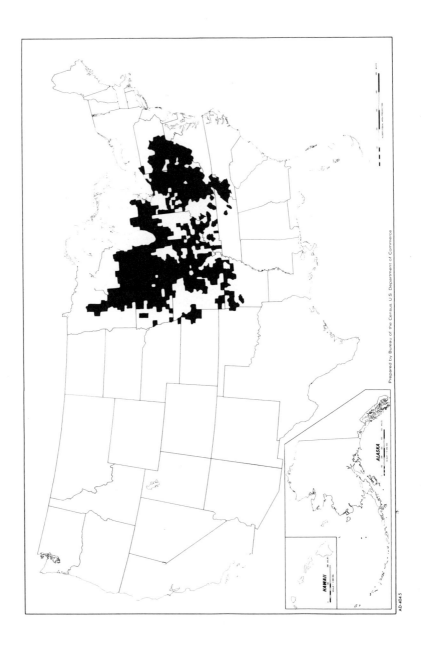

Prepared by Bureau of the Census, U.S. Department of Commerce

AD-404.5

56

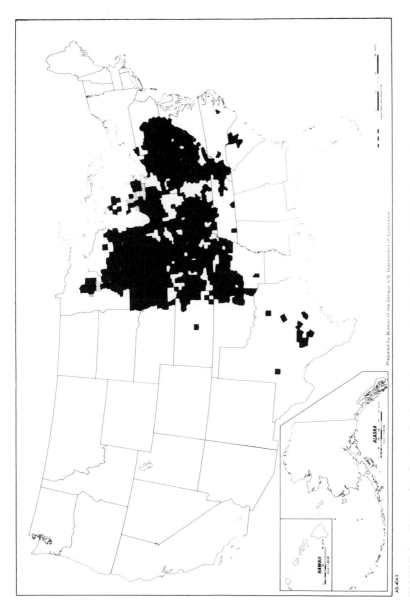

FIGURE 1. Known distribution of oak wilt disease in the United States in 1957 (top), according to True et al. (22) and in 1980 (bottom), according to the U.S. Forest Service (25). The stippled areas on the 1980 map designate counties where the disease was apparently not present in 1980, although it was reported to occur in these areas in 1957.

57

several hypotheses have been tendered as to its recent spread. Mutation, new vectors, or simply increased monitoring and recognition have been mentioned as possible reasons for spread of the disease. Another possibility is that perhaps oak wilt increased *indirectly* because of the chestnut blight. When the chestnut, which comprised 25% of the eastern North American deciduous forest, was removed by blight, red oak apparently increased in density (28,29). Since red oak is susceptible to oak wilt, this increase in its population and the loss of intervening, interspersed chestnuts would render red oak more vulnerable to attack. There would also be more opportunity for root grafting and/or encounters with new vectors with the consequent increase in spread of the disease.

RELATIONSHIP OF COEVOLUTION AND OTHER FACTORS TO BIOCONTROL EFFORTS

How does the foregoing discussion relate to the use of plant pathogens for biocontrol of weeds? It seems too early to determine this at present. The science of using plant pathogens for weed control is too new to provide a basis for writing a "fundamentals" text as Huffaker (30) did for the biological control of weeds with insects. However, it seems that both the classic approach and the biological herbicide approach are useful tactics for specific types of biological weed control (see Chapter 3).

The classic approach is probably more appropriate to rangelands, pastures, or aquatic sites where a minimum of disturbance may favor natural buildup and spread of the biocontrol agent. Indeed, it is in the relatively undisturbed sites that the classical biocontrol of weeds with insects has been most successful for these reasons (30).

One constraint to using imported pathogens is the extreme host specificity required to ensure safety to beneficial species. Thus the classic approach may well be limited to the obligate parasites such as rusts, which often have a high degree of host specificity. However, many of these obligate parasites (2) are "trapped in the coevolutionary rut of host specialization and cause relatively little damage to the host." A notable exception is *Puccinia chondrillina*, which is used to control skeletonweed, *Chondrilla juncea*.

Where disturbance is the norm, as in cultivated fields, the manipulation or augmentation of the weed pathogens as biological herbicides will certainly be required. Pathogens of either native or foreign origin could be used, but use of native pathogens provides several advantages (see Chapters 7 and 8). If an unexpected, undesirable host relationship (disjunction) occurs, the project can be dropped without long-term untoward effects. Moreover, several native pathogens can be investigated concurrently in the same laboratory without quarantine restrictions; however, intentionally introduced exotics must be quarantined. Further, from the standpoint of practicality, we should know which native pathogens are available for use as biological herbicides before spending large sums of money on foreign exploration.

The greatest chance for success in controlling weeds in cultivated fields by the biological herbicide approach would appear to be through manipulation of indigenous pathogens on native plant species or genera closely allied to the targeted exotic weed(s). Thus homeostasis would be minimized. The indigenous pathogen(s) will have to be manipulated to obviate the effects of antagonists, host resistance, and environmental constraints; this can be done (31).

In 1969 Wilson (32) wrote, "The idea of using plant pathogens to control weeds is about as old as the science of plant pathology itself. . . (but) the seeds of the idea have lain dormant since their sowing." This book provides evidence that the dormancy is now broken and that research on the beneficial use of plant pathogens to control weeds has become a dynamic, viable endeavor.

REFERENCES

1. H. D. Ohr. 1974. Plant disease impacts on weeds in the natural ecosystem. *Proc. Am. Phytopathol. Soc.* 1, 181–184.

2. J. L. Harper. 1977. *Population Biology of Plants.* Academic Press, New York, 892 pp.

3. J. J. Burdon and R. C. Shattock. 1980. Disease in plant communities. Pp. 145–219 in: T. H. Coaker, ed., *Applied Biology.* Vol. 5. Academic Press, New York.

4. J. Stein and L. Urdang, eds. 1973. *The Random House Dictionary of the English Language.* Random House, New York, 2059 pp.

5. G. Rangaswami. 1966. *Agricultural Microbiology.* Asia Publishing House, New York, 413 pp.

6. J. E. van der Plank. 1960. Analysis of epidemics. Pp. 229–289 in: J. G. Horsfall and A. E. Dimond, eds., *Plant Pathology: An Advanced Treatise.* Vol. 3. *The Diseased Population: Epidemics and Control.* Academic Press, New York.

7. J. G. Horsfall and E. B. Cowling. 1978. Some epidemics man has known. Pp. 17–32 in: J. G. Horsfall and E. B. Cowling, eds., *Plant Disease: An Advanced Treatise.* Vol. 2. *How Disease Develops in Populations.* Academic Press, New York.

8. G. L. Carefoot and E. R. Sprott. 1967. *Famine on the Wind.* Rand McNally, Chicago, IL, 231 pp.

9. E. G. Kuhlman. 1978. The devastation of American chestnut by blight. Pp. 1–3 in: W. L. MacDonald, F. C. Cech, J. Luchok, and C. Smith, eds., *Proceedings of the American Chestnut Symposium.* West Virginia University, Morgantown, WV.

10. B. K. Hartline. 1980. Fighting the spreading chestnut blight. *Science* 209, 892–893.

11. J. N. Gibbs. 1980. Dutch elm disease and the wicker basket theory. *Phytopathology* 70, 699.

12. P. W. Talboys. 1978. Dysfunction of the water system. Pp. 141–162 in: J. G. Horsfall and E. B. Cowling, eds., *Plant Disease: An Advanced Treatise*. Vol. 3. *How Plants Suffer from Disease*. Academic Press, New York.

13. J. R. Allison. 1978. *How to Identify Dutch Elm Disease*. U.S. Forest Service, Forest Insect and Disease Management, Northeastern Area, State and Private Forestry Publication, Delaware, OH, 2 pp.

14. F. D. Podger, R. F. Doepel, and G. A. Zentmyer. 1965. Association of *Phytophthora cinnamomi* with a disease of *Eucalyptus marginata* forest in western Australia. *Plant Dis. Rep.* **49**, 943–947.

15. G. A. Zentmyer, A. O. Paulus, and R. M. Burns. 1967. *Avocado Root Rot*. California Agricultural Experiment Station Circular 511 (revised), 18 pp.

16. W. A. Campbell and O. L. Copeland, Jr. 1954. *Littleleaf Disease of Shortleaf and Loblolly Pine*. U.S. Deparment of Agriculture Circular 940, 41 pp.

17. G. A. Zentmyer. 1963. Biological control of phytophthora root rot of avocado with alfalfa meal. *Phytopathology* **53**, 1383–1387.

18. J. G. Watts, D. R. Liesner, and D. L. Lindsey. 1977. *Salt Cedar—A Potential Target for Biological Control*. New Mexico Agricultural Experiment Station Bulletin 650, 28 pp.

19. P. C. Quimby, Jr., E. B. Hollingsworth, and R. L. McDonald. 1977. Techniques for greenhouse evaluation of herbicides on saltcedar. *Weed Sci.* **25**, 1–4.

20. E. B. Hollingsworth, P. C. Quimby, Jr., and D. C. Jaramillo. 1979. Control of saltcedar by subsurface placement of herbicides. *J. Range Manage.* **32**, 288–291.

21. M. J. Frezzi. 1942. Death of tamarisk caused by *Botryosphaeria tamaricis* in Corrientes, Argentina. *Rev. Argentina de Agron.* **9**, 100–113. (In *Rev. Appl. Mycol.* 1944. **23**, 301.)

22. R. P. True, H. L. Barnett, C. K. Dorsey, and J. G. Leach. 1960. *Oak Wilt in West Virginia*. West Virginia Agricultural Experiment Station Bulletin 448T, 119 pp.

23. R. L. Anderson and D. D. Skilling. 1955. Oak wilt damage—survey in central Wisconsin. *Lake States For. Exp. Stn. Pap.* **33**, 1–11. (Cited by J. S. Boyce. 1961. *Forest Pathology*, 3rd ed. McGraw–Hill, New York, p. 310.)

24. R. Lewis, Jr. and F. L. Oliviera. 1979. Live oak decline in Texas. *J. Arboric.* **5**, 241–244.

25. Anonymous. 1980. *Distribution of Oak Wilt*. U.S. Forest Service, Forest Insect and Disease Management, Southeastern Area, State and Private Forestry Map, Atlanta, GA, 1 p.

26. F. H. Tainter and W. D. Gubler. 1973. Natural biological control of oak wilt in Arkansas. *Phytopathology* **63**, 1027–1034.

27. E. R. Pianka. 1978. *Evolutionary Ecology*, 2nd ed. Harper and Row, New York, 397 pp.

28. H. J. Oosting. 1956. *The Study of Plant Communities*, 2nd ed. Freeman, San Francisco, 440 pp.

29. J. F. McCormick and R. B. Platt. 1980. Recovery of an Appalachian forest following the chestnut blight or Catherine Keever—you were right! *Am. Midl. Nat.* **104**, 264–273.

30. C. G. Huffaker. 1957. Fundamentals of biological control of weeds. *Hilgardia* **27**, 101–157.

31. G. E. Templeton, D. O. TeBeest, and R. J. Smith, Jr. 1979. Biological weed control with mycoherbicides. *Annu. Rev. Phytopathol.* **17**, 301–310.

32. C. L. Wilson. 1969. Use of plant pathogens in weed control. *Annu. Rev. Phytopathol.* **7**, 411–434.

5

CONSTRAINTS ON DISEASE DEVELOPMENT

G. E. HOLCOMB

*Department of Plant Pathology and Crop Physiology,
Louisiana Agricultural Experiment Station, Louisiana State University,
Baton Rouge, Louisiana*

The ordinary citizen today assumes that science knows what makes the [biotic] community clock tick; the scientist is equally sure that he does not. He knows that the biotic mechanism is so complex that its workings may never be fully understood.

Aldo Leopold, in *A Sand County Almanac*

The advancement of a scientific discipline such as plant pathology depends on the continuous challenge of traditional views and concepts. Definitions of terms are refined and altered. Concepts of the discipline and relationship to sister disciplines are reconsidered. New challenges are put forth by forward-viewing members with the intent to stimulate or renew advancement. Bateman (1) did this with the thought provoking statement that "too many [plant pathology] researchers are continually reinventing the wheel." He then challenged plant pathologists by stating that "plant pathology must embrace more than man's efforts to control plant disease if it is to be of maximum benefit to the society which pays its bills." The contributors to this book are meeting some of these challenges by considering plant pathogens as weed control agents.

TERMINOLOGY AND DEFINITIONS

This chapter presents some of the terminology and concepts of plant pathology as related to the disease complex and to contraints on disease development. The terminology and concepts will form a basis for understanding the more advanced topics that relate to biological control of weeds with plant pathogens.

Plant pathology is the study of *plant diseases*. It includes the study of the causal agents of disease, the manner in which disease is induced, and the interactions between the causal agents and the diseased host plant (2). Concepts of disease in plants have included consideration of injury, disturbances in form and physiological functions, and economic loss to humans. Plant diseases are almost always considered to be detrimental to the plant. However, the observation that plant disease can be beneficial to humans was discussed recently by Wilson (3), who cited weed biocontrol with plant pathogens as a notable example.

A concept of disease was stated in modern terms by Bateman (1): "*Disease* is the injurious alteration of one or more ordered processes of energy utilization in a living system, caused by the continued irritation of a primary causal factor or factors." This definition as applied to *plant disease* implies (1) injury to the plant, (2) an alteration in the orderly use of energy by the plant, and (3) that these conditions are induced by the continued irritation of the causal factors. This eliminates consideration as plant diseases those injuries that result from transient irritations, such as severe weather damage (wind, hail, etc.) or the chewing of a rodent. Injurious alterations caused by a fungus, an air pollutant, or a gall-forming insect are plant diseases.

The factors or agents that cause plant disease are referred to as *pathogens* and may be living (biotic) or nonliving (abiotic). The plant that is being acted on by a pathogen is the *host* or *suscept*, and outward manifestations of disease are called *symptoms*. The *host range* of a pathogen is the kinds of plants a specific pathogen can infect. The infective unit or propagule (e.g., the spore) that is carried to the host plant and initiates infection is referred to as *inoculum*. *Infectious diseases* are those caused by living organisms (e.g., fungi and bacteria) or viruses in contrast to *noninfectious diseases* caused by nonliving pathogens such as air pollutants. Infectious pathogens can be transmitted or carried from a diseased plant to a healthy plant and cause disease in the healthy plant. *Infection* is the process in which a parasite or biotic pathogen establishes contact with the host tissues. *Pathogenicity* is the ability of a pathogen to produce disease in a given host, and *virulence* refers to the degree of pathogenicity. Nelson (4) considered virulence to be the relative amount of disease produced when two or more pathogenic isolates of an organism are compared on a specific host genotype. He used the term "non-pathogenic" instead of "avirulent" for an isolate that does not cause disease. Finally, *pathogenesis* is the orderly process of disease development in the host.

CLASSIFICATION OF PLANT DISEASES

Classification is done to facilitate the study and understanding of plant diseases. One of the most useful, frequently used, and convenient classifications is based on the kinds of causal agents (5). Disease-causing agents can readily be separated into living (infectious) and nonliving (noninfectious) agents (Table 1). Among the living organisms that cause disease are fungi, bacteria, and nematodes. Viruses are placed in this group for convenience because they are obligate parasites and require living host cells for their replication. The causal agent classification method is also useful

TABLE 1. Classification of Plant Diseases Based on the Kind of Causal Agent

Infectious Plant Diseases Caused by
1. Fungi
2. Bacteria
3. Nematodes
4. Viruses and viroids
5. Parasitic plants (mistletoe, witchweed)
6. Mycoplasmas
7. Protozoa
8. Algae
9. Insects (e.g., gall formers)

Noninfectious Plant Diseases Caused by
1. Temperature extremes
2. Extremes in amount of moisture
3. Extremes in amount and duration of light
4. Extremes in soil pH and pH of precipitation
5. Air pollution
6. Nutrient deficiencies
7. Mineral toxicities
8. Lack of oxygen
9. Pesticide toxicities

in the study of diseases on specific groups of plants. Thus a second method can be used to classify diseases as vegetable diseases, cereal diseases, field crop diseases, alfalfa diseases, and so on.

A third method of disease classification is based on the types of symptom expressed on the host plant. Symptom terminology may also indicate the plant part infected, and this terminology has been used in classification. Terms in common usage include leaf spot, blight, canker, root rot, wilt, scab, chlorosis, mildew, stunt, gall, dieback, necrosis, and shothole. These symptoms can be placed into three broad groups as follows: necroses—degeneration leading to death (necrosis) of cells, tissues, and whole plants; hypoplases—failure of parts or the whole plant to develop fully (hypoplasia); and hyperplases—overdevelopment of plants or plant parts (hyperplasia) (6). A compromise classification system was used by Wheeler (7), partly on the basis of major symptoms (root and foot rot, blight, leaf spots, etc.) and partly according to the common names given to major fungal pathogen groups (rust, smut, downy mildew, and powdery mildew).

A fourth and currently popular classification of plant disease is based on the disturbances of vital plant processes caused by pathogens (8). Pathogen interference with these processes leads to the following six classes of pathogenesis: (1) tissue disintegration; (2) effects on growth; (3) effects on reproduction; (4) effects on nutrition; (5) effects on water balance; and (6) alteration in respiratory patterns. A seventh class was recently added to this list, namely, the pathogen interference in energy capture and use by the host plant (9). This approach to the study of disease is quite useful because it emphasizes what happens to the plant in terms of its vital physiological processes.

PARASITISM AND DISEASE

Parasitic associations are those in which one organism (the parasite) benefits by living in or on another organism (the host) and obtains its nutrition from the host. Since a parasitic relationship is not necessarily harmful to the host, a distinction is made between parasitism and disease, with the latter harmful to the host (suscept). Parasitic relationships are divided into (1) those in which both parasite and host benefit (mutualistic or symbiotic), (2) those in which usually only the parasite benefits, but without harm to the host (commensalistic), and (3) those in which only the host or both host and parasite are injured or killed (pathosistic) (1). The point is made that not all parasites are pathogens and not all biotic pathogens are parasites. (An example of the latter is the fungus *Sclerotinia* that produces toxins and enzymes that kill host tissues in advance of fungal invasion.) Plant pathology is concerned primarily with parasitic relationships that result in disease (pathosism).

Biotic plant pathogens have traditionally been classified as (1) obligate parasites (those incapable of living apart from their hosts), (2) facultative parasites (those with mainly a saprophytic habit but have weak parasitic properties), and (3) facultative saprophytes (those in which the parasitic habit is strongly developed but can exist as saprophytes for a part of their life cycle) (5). Saprophytes are on the opposite end of the food spectrum from parasites and obtain food only from dead organic matter. Brian (10) divided obligate fungal parasites into (1) those ecologically obligate and (2) those physiologically obligate that are unable to survive saprophytically in pure culture even on complex media.

Luttrell (11) placed plant parasitic fungi into three broad categories based on "mode of nutrition and way of life" as an aspect of fungal ecology (other parasitic organisms can also be classified in this manner):

1. *Biotrophs.* Pathogens such as smut and rust fungi that obtain their food in nature only from living tissue (regardless of the ease of culturing in the laboratory).

2. *Hemibiotrophs.* Typified by leaf-spotting fungal pathogens that infect living tissue in the same manner as biotrophs but continue to develop and sporulate after the host tissue dies.

3. *Perthotrophs.* Pathogenic fungi, such as *Armillariella mellea*, that kill host tissue in advance of penetration with freshly killed tissue serving as a food base for further advancement. These fungi grow saprophytically from sclerotia or food bases and are essentially saprophytes with the ability to kill the tissues on which they live.

An understanding of pathogenesis is central to the understanding of plant pathology. Attempts to understand disease in terms of parasite nutrition and loss of nutrients by the host does not account for all host injury. The attacking mechanism of the pathogen usually accounts for more injury than does the loss of nutrients (1). Bateman (1) proposed a "multiple-component" hypothesis of pathogenesis and parasitism in which the pathogen and host are considered to present both a favorable and an unfavorable environment to each other. Any given environment is made up of multiple factors, such as nutrients, toxins, enzymes, and growth regulators. Bateman (1) stated, "It is the sum of the components of the four environments and their interaction that determines the type of host-parasite relationship—mutualistic, commensalistic, or pathosistic."

THE DISEASE TRIANGLE

Three important components of the disease complex that are necessary for disease to occur are a susceptible host, a virulent pathogen, and favorable environmental conditions. The "disease triangle" is often used to illustrate the complex relationships and interactions among these three components (Fig. 1a). A fourth major component, time, must now be considered as a part of the disease complex, although a sound understanding of its role was established only about 30 years ago. Browning et al. (12) incorporated the time component into a geometric diagram that they named the "disease pyramid." The three baselines of a three-faced pyramid represent the original components in the disease triangle, and the time component is depicted as the internal volume of the pyramid and represented by a broken line drawn from the apex to the base. This complex interrelationship of components can also be shown by drawing a time component circle inside the familiar disease triangle (Fig. 1b).

A broadened or expanded concept of the disease triangle, encompassing the four major components, comes from the study of disease epidemics. The 1970 southern corn leaf blight epidemic in the United States

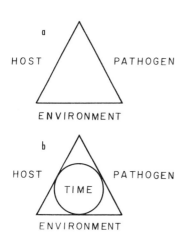

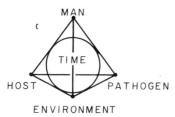

FIGURE 1. Diagramatic representation of interactions between components of the disease complex: (a) plant disease triangle; (b) plant disease triangle incorporating the time component; (c) disease component diagram in which human influence on disease is recognized.

presented dramatic evidence of what happens when all disease components and factors interact favorably for disease development. Zadoks and Schein (13) emphasized human influence on disease development and illustrated human interaction with the original three disease components (but omitted time) in the form of a "disease tetrahedron." This concept is illustrated in Fig. 1c, in which the time component has been added. Cowling (14) discussed human influence on disease development in terms of management practices that favor epidemics.

CONSTRAINTS ON DISEASE DEVELOPMENT

Constraints on disease development are those factors that retard or prevent disease development (2,5,14–17). They include obvious factors such as environmental temperatures unfavorable to either or both host and pathogen and less obvious factors, such as genetic makeup of host and pathogen populations. Factors that act as constraints to disease development are considered in relation to their effects on the major components of the disease complex, especially the host and the pathogen (Table 2).

HOST-RELATED CONSTRAINTS

One of the most important host-related constraints to disease development is resistance to disease. However, the dramatic destructiveness of some pathogens on highly susceptible, cultivated plants tends to obscure the fact that most plants are resistant to most pathogens. Two other important host constraints are the relative abundance of the host and its geographic range and distribution. Mixed plantings offer constraints to the rapid spread of disease that are not present in pure stands. Diversity of the host genetic base increases the possibility of resistance appearing and thus serves as a constraint to disease (18). An example of disease destruction that can occur when these constraints are not present is Dutch elm disease. This disease has destroyed large numbers of American elm that were planted as principal shade trees in many cities of the northeastern and northcentral United States. The lack of resistance and the genetic uniformity of the elm populations contributed greatly to the "success" of this disease. See Chapter 4 for further information on Dutch elm disease.

Other host-related factors that may help constrain disease development include age and vigor of the host plant. It is widely recognized that succulent seedlings are susceptible to infection and damage by damping-off fungi but become essentially immune to attack when tissues mature and harden. The opposite of this response is also common for certain plant-

TABLE 2.   Factors That May Act as Constraints to Disease Development

---

Host-Related Factors

1. Degree of host resistance or susceptibility to a pathogen
2. Physical and physiological barriers to infection in the host
3. Abundance and distribution of host
4. Age and vigor of host
5. Predisposition of host by environmental factors
6. Host effects on environment
7. Diversity of host's genetic base

Pathogen-Related Factors

1. Host range
2. Virulence
3. Types and numbers of propagules produced by the pathogen
4. Mechanisms for genetic variability of the pathogen
5. Vector relationships and other means of pathogen dissemination
6. Means of pathogen survival from season to season

Environmental Factors that Affect Host and Pathogen

1. Temperature and wind
2. Moisture (precipitation and humidity)
3. Form and availability of nutrients
4. Radiation
5. Oxygen, carbon dioxide, and pH levels in the soil

Time Factor

1. Duration as it affects sequence and interactions

Management (human) practices that affect host and pathogen

---

pathogen combinations in which the immature plants or plant parts are resistant, but the mature plants or plant parts are susceptible to certain pathogens. Examples of this age-related resistance is that of stone fruits to *Monilinia* (brown rot) and many leaf spot diseases caused by *Cercospora*. Plants maintained in a vigorous growing condition are generally less susceptible to many pathogens, although vigorous growth resulting from excess nitrogen fertilizer can increase a plant's susceptibility to some diseases such as fire blight of pear (2,20).

ENVIRONMENTAL CONSTRAINTS

Environmentally conditioned susceptibility, referred to as *predisposition*, results from the effects of nongenetic conditions acting before infection to alter the resistance or susceptibility of plants to disease (19,20). Environmental factors, such as temperature, moisture, light, nutrient levels, and pH may influence predisposition and thus may act as constraints to disease development. Environmental factors (Table 2) also directly affect the interactions of host and pathogen, and their presence, absence, amount, and duration may act to constrain disease (21).

PATHOGEN-RELATED CONSTRAINTS

Consideration of pathogen-related constaints on disease development can be approached by reviewing some characteristics of a successful pathogen (15). These characteristics include (1) production of large numbers of propagules, (2) high survival capacity of propagules, (3) rapid germination and growth of propagules, (4) antibiotic production by propagules to inhibit antagonists and competitors, (5) capacity for rapid infection and invasion of the host, (6) capacity to survive adverse environmental conditions, (7) ability for saprophytic growth, (8) ability to parasitize without causing great host damage, (9) wide host range, and (10) genetic capacity to adapt to genetic variations of the host. The lack of any of these characteristics in a pathogen renders it less competitive and thus serves as a constraint. All pathogen-related and environmental factors (Table 2) that adversely affect growth, reproduction, dissemination, parasitic and pathogenic capacities, and survival act as constraints to disease development.

TIME AS A CONSTRAINT

Time as a factor in disease development influences all components independently and also the complex interactions between these components. Time is an important factor and influences disease development in many ways, such as the time of year during which the host and pathogen are available to interact and the time and duration of moisture and temperature conditions favorable to both host and pathogen. Unfavorable duration of interaction or exposure to certain environmental factors would act as constaints to disease development.

The complexity of interactions that occur during the development of an epidemic was emphasized by Day (18). He indicated that for 19 variables (five host-related, six pathogen-related, and eight related to the environment) there are 361 possible combinations of interactions.

HUMAN INFLUENCE AS A CONSTRAINT

Human influence on the plant disease complex has also been emphasized in relation to epidemiology and disease management (13–15). Many well-

TABLE 3. Management Practices That Serve as Constraints to Disease Development

1. Breeding for disease resistance but maintaining genetic diversity
2. Use of crop rotations that reduce pathogen populations including trap crops
3. Soil amendments that inhibit pathogens or stimulate their antagonists
4. Vector control
5. Changing soil pH to favor antagonists and/or inhibit pathogens
6. Use of tillage methods that reduce pathogen populations
7. Avoiding crowding in host propagation nurseries
8. Use of irrigation practices that reduce or prevent pathogen reproduction and survival
9. Production of disease-free propagating materials
10. Selection of planting dates that are unfavorable to pathogens
11. Use of chemical treatments that reduce pathogen populations or stimulate antagonists
12. Harvesting, handling, and storage practices that avoid or reduce wounding

known epidemics of modern times can be attributed directly to human activities. Management practices, however, have been developed to reduce disease impact. They include the powerful tools of breeding for resistance and the use of chemicals that inhibit and/or destroy pathogens (Table 3). Management of any factor that affects disease so as to favor the host— namely, host susceptibility, chemical treatments, crop rotation, planting dates, and so on—will reduce disease and serve as a constraint to disease development.

Many of the factors that are considered constraints to disease development may be manipulated or used in a manner that will enhance development of disease. Age-related susceptibility of a weed host, moisture and temperature conditions that favor disease development, and selection for virulence in weed pathogens are examples of factors that can be used to increase the effectiveness of plant pathogens that are being evaluated as weed control agents.

In conclusion, it is the goal of plant pathology to understand the disease complex so that plant diseases can be managed or controlled. This same knowledge, however, can be used to manipulate pathogens for the biological control of weeds.

REFERENCES

1. D. F. Bateman. 1978. The dynamic nature of disease. Pp. 53–83 in: J. G. Horsfall and E. B. Cowling, eds., *Plant Disease: An advanced Treatise*. Vol. 3. *How Plants Suffer from Disease*. Academic Press, New York.

2. G. N. Agrios. 1978. *Plant Pathology*, 2nd. ed. Academic Press, New York, 703 pp.

3. C. L. Wilson. 1977. Management of beneficial plant diseases. Pp. 347–362 in: J. G. Horsfall and E. B. Cowling, eds., *Plant Disease: An Advanced Treatise*. Vol. 1. *How Disease is Managed*. Academic Press, New York.

4. R. R. Nelson. 1973. Pathogen variation and host resistance. Pp. 40–48 in: R. R. Nelson, ed., *Breeding Plants for Disease Resistance*. Pennsylvania State University Press, University Park, PA.

5. J. C. Walker. 1969. *Plant Pathology*, 3rd ed. McGraw-Hill, New York, 819 pp.

6. D. A. Roberts and C. W. Boothroyd. 1972. *Fundamentals of Plant Pathology*. Freeman, San Francisco, 424 pp.

7. B. E. J. Wheeler. 1969. *An Introduction to Plant Diseases*. Wiley, New York, 374 pp.

8. G. L. McNew. 1960. The nature, origin, and evolution of parasitism. Pp. 19–69 in: J. G. Horsfall and A. E. Dimond, eds., *Plant Pathology: An Advanced Treatise*. Vol. 2, *How Disease Develops in Populations*. Academic Press, New York.

9. T. Kosuge. 1978. The capture and use of energy by diseased plants. Pp. 85–116 in: J. G. Horsfall and E. B. Cowling, eds., *Plant Disease: An Advanced Treatise*. Vol. 3. *How Plants Suffer from Disease*. Academic Press, New York.

10. P. W. Brian. 1967. Obligate parasitism in fungi. *Proc. R. Soc. B* **168**, 101–118.

11. E. S. Luttrell. 1974. Parasitism of fungi on vascular plants. *Mycologia* **66**, 1–15.

12. J. A. Browning, M. D. Simons, and E. Torres. 1977. Managing host genes: epidemiologic and genetic concepts. Pp. 191–212 in: J. G. Horsfall and E. B. Cowling, eds., *Plant Disease: An Advanced Treatise*. Vol. 1. *How Disease is Managed*. Academic Press, New York.

13. J. C. Zadoks and R. D. Schein. 1979. *Epidemiology and Plant Disease Management*. Oxford University Press, New York, 427 pp.

14. E. B. Cowling. 1978. Agricultural and forest practices that favor epidemics. Pp. 361–381 in: J. G. Horsfall and E. B. Cowling, eds., *Plant Disease: An Advanced Treatise*. Vol. 2. *How Disease Develops in Populations*. Academic Press, New York.

15. K. F. Baker and R. J. Cook. 1974. *Biological Control of Plant Pathogens*. Freeman, San Francisco, 433 pp.

16. N. E. Stevens and R. B. Stevens. 1952. *Disease in Plants*. Chronica Bontanica Company, Waltham, MA, 219 pp.

17. R. B. Stevens. 1960. Cultural practices in disease control. Pp. 357–429 in: J. G. Horsfall and A. E. Dimond, eds., *Plant Pathology: An Advanced Treatise*. Vol. 3. *The Diseased Population: Epidemics and Control*. Academic Press, New York.

18. P. R. Day. 1974. *Genetics of Host-Parasite Interaction*. Freeman, San Francisco, 238 pp.

19. J. Colhoun. 1979. Predisposition by the environment. Pp. 75–96 in: J. G. Horsfall and E. B. Cowling, eds., *Plant Disease: An Advanced Treatise*. Vol. 4. *How Pathogens Induce Disease*. Academic Press, New York.

20. C. E. Yarwood. 1959. Predisposition, Pp. 521–562 in: J. G. Horsfall and A. E. Dimond, eds., *Plant Pathology: An Advanced Treatise*. Vol. 1. *The Diseased Plant*. Academic Press, New York.

21. J. Colhoun. 1973. Effects of environmental factors on plant disease. *Annu. Rev. Phytopathol.* **11**, 343–364.

6

GENETIC VARIATION IN WEEDS

S. C. H. BARRETT
Department of Botany, University of Toronto, Toronto, Ontario, Canada

Evolution by natural selection depends on the presence of heritable variation in populations. Fisher (1) demonstrated that there is a precise relationship between genetic variation and the maximum possible rate of natural selection. Hence the rate of increase in fitness of a population is equal to its genetic variance in fitness at that time. Therefore, it is of some importance in evolutionary studies to determine the amount and nature of genetic variation present in natural populations of organisms. In fact, the study of genetic variation has become a major preoccupation of evolutionary biologists in recent years (2–4).

Weedy plant species are suitable organisms for genetic and evolutionary studies because they are usually abundant, grow rapidly, and produce large numbers of offspring (5). As a result, some information on the genetic structure and levels of variation in populations of weed species is now available. The objective of this chapter is to review the existing literature concerned with genetic variation in weeds and to discuss some of the factors that determine the genetic structure of populations of colonizing plants. An attempt is made to relate this information to the methods employed in biological weed control.

WEED CHARACTERISTICS

There are many different definitions of what constitutes a weed (6–8). For the purpose of this discussion, I shall use the botanical definition of Baker

(9) that a plant is a weed, "if in any specified geographic area, its populations grow entirely or predominantly in situations markedly disturbed by man, without being deliberately cultivated." I shall also refer to particular weed species as ruderals, which occur primarily on waste ground and along roadsides, and agrestals, which are found principally as weeds of agroecosystems. These latter categories are not mutually exclusive since many weeds can occur in either situation. Aquatic weeds comprise a third class of weeds that is usually restricted to canals, drainage ditches, reservoirs, and other forms of water impoundment.

Although weed species have evolved independently in many diverse taxonomic groups, they frequently share similar sets of adaptations or life history traits. It is worthwhile to consider some of these attributes since they strongly influence the genetic properties of weeds and hence their evolutionary potential. Baker (5,9) has compiled a list of "ideal" weed characteristics. Some common features of weed species are presented in Table 1. Not all weeds exhibit these traits, and many nonweeds possess some of these characteristics. Nevertheless, it is useful to keep these features of the life history of weeds in mind when analyzing individual cases of colonizing success and in comparative studies of related weed and nonweed groups (5, 9–11).

In a review of the biological characteristics of the world's 18 most serious weeds, Brown and Marshall (12) identified four common features: (1) reproduction either by self-fertilization or clonal propagation with a high reproductive capacity; (2) extensive continental distribution and hence adaptation to a wide range of environments; (3) ecotype formation; and (4) polyploidy. These characteristics all have important genetic implications and are useful in formulating a general theory for the population biology and evolution of weeds.

SELECTION PRESSURES ON WEEDS

Many weed species have survived and evolved over many centuries of environmental change (13). This is particularly notable in Eurasia, where there is a long history of large-scale disturbances through agriculture and deforestation. Many of the most widespread and successful weeds in North America are Eurasian in origin and have been introduced to the New World in historic times (14,15). One of the reasons for the success of these aliens over native North American species may be related to their longer association with agriculture. This longer association provides time

TABLE 1. Some Common Features of Weeds

1. Dormancy and asynchronous germination
2. Rapid development
3. Phenotypic plasticity
4. Clonal, apomictic, or autogamous reproduction
5. High reproductive capacity
6. Small, easily dispersed seeds
7. Ecological race formation
8. Polyploidy

for the evolution of characteristics that are favored under disturbance (16).

Improvements in agricultural techniques result in altered selection pressures on weed populations. These forces are often of a general nature affecting entire weed floras. The introduction of new crops, as well as alterations in fertilizer treatments, times of sowing, herbicide usage, and crop spacing will influence most weed species in some manner. For example, the development of continuous flooding techniques in Californian rice fields has altered the composition of the weed flora by reducing infestations of weed taxa such as *Ammannia coccinea*, *Echinochloa crusgalli* var. *crus-galli*, and *Leptochloa fascicularis*, which are unable to establish in deep water (17). In some cases the selective forces are highly specific, as with the use of biological control methods intended to reduce populations of particular weed species (18). Responses to these selection pressures vary. Some species are unable to adapt and are replaced by other weed taxa. In California *E. crus-galli* var. *oryzicola*, a large seeded variety of barnyardgrass that is capable of establishing in deep water, has now replaced *E. crus-galli* var. *crus-galli* as the most serious weed of rice in the state (17,19). Alternatively, if the appropriate genetic variation is available, populations may respond by evolving new adapted forms, as has occurred in several weed species subjected to regular applications of S-triazine herbicides (discussed later in this chapter). The important point is that natural selection is an ongoing process in today's agroecosystems and weed adaptation will depend on the strength of the selective forces applied by the agriculturalist as well as the genetic composition of weed populations. At this time, attempts to predict the outcome of these interactions is a difficult task because of our general ignorance of the adaptive value of most forms of genetic variation (20).

THE NATURE OF VARIATION

PHENOTYPIC PLASTICITY

Populations of weeds often exhibit striking variation in size and expression of morphologic characters. Much of this variability is a consequence of phenotypic plasticity: the ability of individual genotypes to alter their growth and development in response to changes in environmental factors. Several authors (9,12, 21–23) have discussed the ecological importance of phenotype plasticity and its adaptive significance to weed populations.

By growing plants of *Echinochloa crus-galli* var. *crus-galli* from a single population under stressed and fertilized conditions, it is possible to produce plants which vary by as much as 8537 times in above-ground biomass and 9410 times in seed fecundity (Table 2). This type of environmentally induced variation presumably enables genotypes of barnyardgrass to survive and reproduce in the heterogeneous and unpredictable environments associated with seasonally flooded land.

Many strictly aquatic weeds also exhibit striking phenotypic plasticity. In a single colony, plants of *Eichhornia crassipes* (waterhyacinth) may differ markedly in size, leaf shape, and petiole swelling. Such diverse phenotypes, which result from different light and nutrient conditions, have been treated as ecotypes (24) or confused with related species such as

TABLE 2.   Phenotypic Plasticity in Echinochloa crus-galli **var.** crus-galli **(Barnyardgrass)** [a]

Character Measured ($n = 10$) [b]	Fertilized Treatment	Density-Stress Treatment
Above-ground biomass (g)	219.4	0.0257
Time to flowering (days)	54.1	95.0
Tiller production	17.6	1.0
Seed production	17,880.0	1.9
Harvest index (%)	13.4	10.4

Source: Ref. 11 and unpublished data.

[a] Seed was collected from a population at the periphery of a rice field in California. Plants in the fertilized treatment (NPK) were grown singly in 31,860-cm^3 pots; plants in the density-stress treatment were grown at a starting density of 100 plants per 88-cm^3 pot. Both treatments were set up during May–September 1978 under glasshouse conditions.

[b] Average values.

Eichhornia azurea (examination of major herbarium collections of waterhyacinth confirms this point). In fact, the contrasting forms may often be members of the same clone. Thus the great morphologic variation observed in weed populations under field conditions may not necessarily reflect a high level of genetic diversity.

In California the two annual wild oats *Avena barbata* and *A. fatua* exhibit contrasting patterns of phenotypic and genetic variability. *Avena barbata* is less variable genetically than *A. fatua* but exhibits greater overall phenotypic variation in natural populations. It appears that *A. barbata* relies less on genetic diversity and more on phenotypic plasticity than does *A. fatua* in adapting to heterogeneous environments (23,25,26).

GENETIC DIFFERENTIATION

Variation in weed species also results from genetic differentiation, both within and between populations. The extent of intraspecific variation varies widely among species. In some groups the variability is treated taxonomically by the formal naming of subspecies or varieties. More often the variants are simply recognized as ecotypes or ecological races. Race formation has been described in many common weeds; see Baker (5,9) and Holm et al. (27) for specific examples. Discussions of the systematic and evolutionary implications of intraspecific variation in plant populations are available in Davis and Heywood (28) and Heslop-Harrison (29). As might be anticipated, the evolution of genetic differentiation in weed species is favored if populations can persist in an area long enough to enable adaptation to the local environment.

THE MEASUREMENT OF VARIATION

Several measures can be used to quantify the levels of genetic variation in a species. Genetic diversity resides both within and between individuals, populations, and regions. Marshall and Brown (30) recognized two major classes of measurement: (1) measures based on genetic variance in quantitative traits such as plant height, time to flowering, and total biomass and (2) measures of allelic diversity at loci governing qualitative characters. These are usually genetic polymorphisms for observable morphologic and biochemical characters.

Measures of genetic diversity based on quantitative characters involve the collection of progenies from individual genotypes from the field. Random sampling of seed parents both within and among populations is usually employed. Families of 5–20 genotypes are grown from each parent in replicated experiments under uniform conditions and measured for a range of vegetative and reproductive parameters. Statistical comparisons are made of the partitioning of phenotypic variance in metrical characters both within and among families and populations. These techniques have been used extensively by population biologists interested in measuring genetic variation in colonizing species (10,25,31–35), as well as by plant breeders analyzing the components of yield (36). However, estimates of variation in quantitative characters are merely indirect measures of genetic diversity since they reflect only that portion of the genetic variability that is expressed phenotypically. The amount of variability actually observed will depend on the character being measured as well as the genetic background and environmental conditions in which it is expressed (30).

Application of the techniques of gel electrophoresis to population genetic studies during the past 15 years has enabled a more direct assessment of the levels of genetic variation in plant and animal populations. The degree of genic variation in a species can be estimated if allelic variants of single genes representing a random sample of the total genome can be detected (37). Electrophoretic techniques permit the detection of allelic variants at enzyme loci. Since most enzymes are the products of individual genes, it is possible, to a first approximation, to equate enzyme variation with variation in genes. Variant and invariant gene loci can be identified and a random sample of genes surveyed. Reviews of electrophoretic techniques and their application to the measurement of genetic variation include those of Lewontin (20), Ayala (38), and Brown (39).

Not all allelic variants are detected by electrophoresis; consequently, the amount of genetic variation is underestimated (20,40). However, at present there is no evidence to indicate that there is a bias in the detection of variability associated with particular ecological or taxonomic groups. Hence the underestimate of allelic variability is probably not a major problem in comparative studies (4).

The basic data obtained from electrophoretic surveys of enzyme and protein variation consist of the genotypic or allelic frequencies at each locus sampled within a population. Four measures of intrapopulation variation are commonly calculated: the mean number of alleles per locus (A); the percentage of loci that are polymorphic (P); a polymorphic index equivalent to the heterozygote frequency under Hardy-Weinberg equilibrium (PI); and the mean heterozygosity per individual (H).

Since the first application of electrophoretic techniques to population biology, over 100 different plant species have been assayed for levels of genetic variation including some 20–30 weed species. Several reviews of the plant data are available. Gottlieb (41) and Brown (39) discussed the relationship between the mating systems of populations and genetic variation, and Gottlieb (42) examined the biochemical consequences of the speciation process. Hamrick (43) and Hamrick et al. (4), while summarizing data from electrophoretic surveys of 113 plant taxa, examined the relationship between life history parameters and levels of genetic variation. Much of the following discussion utilizes data from these reviews.

FACTORS THAT INFLUENCE GENETIC VARIATION

Many factors determine the genetic structure and levels of variation present in a population at a given time. These include the factors that regulate recombination, such as chromosome number, frequency of crossing over, sterility and fertility barriers, mating system, pollination system, dispersal system, and the life history strategy of the species. In addition, various historical and ecological factors such as founder effect, genetic drift, and the degree of environmental heterogeneity can also play a major role. It is an extremely difficult task to determine the extent to which these factors are acting separately or in concert to affect variability. Nevertheless, some patterns are emerging from comparative studies of closely related species as well as from general surveys. This review focuses on those factors that are most relevant to the population genetics of colonizing species and for which data are available (Table 3).

TABLE 3. Some Factors That Influence the Levels of Genetic Variation in Weed Populations

Factor	Influences
Founder effect	Numbers and types of immigrants
Genetic drift	Frequency of genetic bottlenecks during colonizing episodes
Population age and size	Stability of habitat and life history features of species
Degree of sexuality	Importance of cloning and frequency of sexual reproduction
Mating system	Level of inbreeding
Hybridization	Opportunities for gene exchange with related taxa
Environmental heterogeneity	Spatial and temporal variation of habitat

HISTORICAL FACTORS

In many regions of the world and in many crops, the weed flora is composed primarily of alien or introduced species (14,44,45), although exceptions do occur (17). Successful colonization of new territory by introduced species will depend on the degree of preadaptation to the new environment, the number of immigrant propagules that arrive, and the availability of suitable habitats. In many cases of continental migration, only a part of the genetic variation present in a species will be transferred to the new area of occupation. Consequently, populations of alien weeds, particularly those that originate from a single or limited number of introductions, may contain low genetic variation as a result of such "founder effects" (46).

Founder effects combined with an absence or restriction of sexual reproduction can result in striking cases of genetic uniformity. The dioecious aquatic weeds *Elodea canadensis* in Europe (47) and *Myriophyllum brasiliense* in California (14) serve as examples. In the absence of male flowers in the areas of introduction of both species, plants can reproduce only by clonal propagation. Nevertheless, considerable range extensions have been achieved as in the case of *E. canadensis*, which has become a serious weed problem throughout the inland waterways of Europe. It is possible that many widespread aquatic weeds, such as *Salvinia molesta* (48,49), *Hydrilla verticillata* (50), and *Eichhornia crassipes* (51–53), which reproduce predominantly by clonal means, are represented by relatively few genotypes in parts of their alien ranges.

Not all populations of alien weeds exhibit low variation in comparison with source populations. Brown and Marshall (12) found no differences in the average levels of genetic diversity (measured electrophoretically) between native and introduced populations of *Bromus mollis*, respectively, in England and Australia. In fact, the Australian populations exhibited higher levels of heterozygosity than did those from England. Clearly, founder effects may or may not be important in influencing the levels of genetic variation in weed species. Much will depend on the amount of variability in the source population, the extent of immigration, and the reproductive system of the colonist.

REPRODUCTIVE SYSTEMS

Higher plants exhibit a great diversity of reproductive systems (for reviews, see Refs. 54–56). From the standpoint of their influence on genetic diversity, two extremes can be identified. Populations of some species that often reproduce primarily by cloning, such as the bracken fern *Pteridium aquilinum* and the grasses *Festuca rubra* and *Holcus mollis*, can consist of one or a few clones covering large areas (57). At the other extreme are completely sexual, obligate outbreeders with populations containing large amounts of genetic diversity. Among this group are the long-lived tree species, some of which display the highest levels of genetic variation recorded for any living organism (43). For the purpose of this discussion, four modal classes of reproductive system are treated. They are connected by many intermediate conditions.

Asexual Reproduction

Clonal Propagation. Many perennial weeds are capable of clonal or vegetative propagation in addition to reproduction by seed. Methods include propagation by surface stolons and runners, underground rhizomes, tubers, offset buds, corms and bulbs, adventitious buds on cut stems or fallen leaves, and vegetative propagules arising within a flower or inflorescence (vivipary). In every case, genotype duplication occurs and limited genetic diversity may result with extensive cloning and restricted sexual reproduction. In certain weedy forms of *Oxalis pes-caprae* (9) and *Euphorbia cyparissias* (58,59), all reproduction and spread are by clonal methods because of the genetic sterility of populations.

Genetic monotony probably occurs in parts of the adventive range of the waterhyacinth, which is capable of prolific clonal growth through stolon formation. Although seeds are produced in most populations, they frequently fail to produce plants because of the absence of suitable environmental conditions for germination and seedling establishment (52,53). Evidence to support this view comes from genetic studies and observations of the reproductive biology of populations. Waterhyacinth exhibits the rare genetic polymorphism, tristyly (51). Populations of tristylous species usually contain three genetic forms or morphs that differ from one another in style and stamen lengths. In the Amazon Basin the native range of waterhyacinth, long-, mid-, and short-styled forms occur, whereas in many parts of the adventive range of the species only mid-styled forms occur. This suggests that limited genetic diversity was introduced from the native range.

In California, only the mid-styled form of waterhyacinth occurs. However, open-pollinated seeds from a population at Stockton, California, when grown to flowering under glasshouse conditions segregated long- and mid-styled plants. Absence of the long-styled morph under field conditions in conjunction with observations of an absence of seedlings at the site strongly support the view that cloning is the only method by which the population reproduces (53).

A quite different situation is evident in Florida and Louisiana, where the drainage of water ("drawdown") from infested water bodies is often used for control of certain aquatic weeds. This technique favors sexual reproduction and the buildup of genetic diversity in waterhyacinth populations by providing suitable conditions for seed germination and seedling establishment. The water level fluctuations that occur during the drainage and refilling of water bodies parallel the natural flooding regimes of Amazonian habitats to which waterhyacinth is highly adapted. A range of genotypes, including long-styled and heterozygous and homozygous mid-styled forms as well as flower color variants, occur in the southeastern United States (see Refs. 51, 52, and Barrett, unpublished data). Although some of these forms may have arisen from separate introductions from the Neotropics, it is clear that weed control practices that enhance sexual reproduction favor the buildup of genetic diversity in the region.

If clonal species persist in a suitable habitat for a long period of time, considerable levels of genetic diversity can develop (60–62). This is aided by the fact that clonal species are frequently outbreeders. Nevertheless, it seems likely that in many clonally propagated weeds of agricultural land, periodic habitat destruction will reduce opportunities for the development of high genetic diversity, particularly if sexual reproduction is infrequent.

Apomixis.   Apomixis is the production of viable seeds without fertilization. There are several types of apomictic reproduction, and the embryologic details are complex and may vary in every case (63). The usual result of apomixis is the formation of seeds that are genotypically identical to the maternal parent. Apomixis is widespread in higher plants and is frequent among perennial, polyploid species, including many weeds such as *Chondrilla juncea* (64), *Cortaderia jubata* (65), *Eupatorium adenophorum* (9), *Hypericum perforatum* (59), and *Taraxacum officinale* (66).

In some plant groups apomixis replaces sexual reproduction altogether (obligate apomixis), whereas in others, some seeds are formed by the sexual process and the remainder are produced asexually (facultative apomixis). In apomictic species, population variation is split into a series of homogeneous groupings, each differing from one another in minor features. Facultative apomixis provides enormous breeding and evolutionary potential, enabling both genetic recombination and the replication of successful genotypes (67).

Unfortunately, there have been few detailed studies of genetic variation in apomictic species. Solbrig and Simpson (68) demonstrated the presence of a minimum of four electrophoretically distinct genotypes of

Taraxacum officinale (common dandelion) among 300 plants sampled from three adjacent but contrasting habitats in Michigan. Two of these genotypes were common but differed in their relative abundance at the sites. Detailed uniform garden and transplant studies provided strong evidence that the two biotypes possessed contrasting suites of life history traits and were differentially adapted to their respective habitats (68,69). It is not known whether the adaptive differentiation evolved at the site or resulted from introduction of preadapted weed forms.

Usberti and Jain (70) compared allozyme variation at eight gene loci in six populations of the facultative apomict *Panicum maximum*, a roadside weed in many parts of the tropics (71). Three populations were sexual, and the remainder reproduced by a mixture of sexual and apomictic reproduction. The entirely sexual populations exhibited consistently higher genetic variation within populations with P = 100.0, 50.0; A = 2.69, 1.92; and PI = 0.377, 0.139 (refer to the section on measurement of variation), for the sexual and asexual populations, respectively. However, the asexual populations displayed a considerable degree of genetic heterogeneity among the populations sampled.

More information is required on the genetic structure of populations of apomictic weeds, and it is premature to attempt generalizations. Of crucial importance is the amount of genetic recombination that occurs within populations. The balance between sexual and asexual reproduction has a genetic basis in some apomictic species but can also be influenced by environmental factors (63,67,72).

Sexual Reproduction

Inbreeders. Colonization of unoccupied sites is an important feature of the life history of weeds. Therefore, it is perhaps not surprising that most are capable of self-fertilization and hence autogamous seed production. Autogamy ensures the production of seeds irrespective of the proximity of mates (73), the high genetic fidelity of progeny (74), and the maintenance of adaptive gene combinations (39,75). Many of the most successful and widespread weeds, including *Ammannia coccinea, Avena fatua, Chenopodium album, Echinochloa crus-galli, Erodium cicutarium, Hordeum murinum, Lactuca serriola, Monochoria vaginalis, Oxalis corniculata, Portulaca oleracea, Rottboellia exaltata,* and *Rumex crispus,* are primarily self-fertilizing (5,12,17,31).

Two major viewpoints on the genetic structure of self-fertilizing species have developed during the past three decades. Darlington and Mather (76) and Stebbins (74) argued that populations of inbreeders are likely to be composed of one or a few highly fit homozygotes and thus near uniform populations. On the other hand, in a review of the genetics of inbreeding populations, Allard et al. (77) concluded that although inbreeding should theoretically lead to homozygosis and a greater chance of selection among homozygotes (free variability), this may or may not be realized in natural populations of autogamous plants. After reviewing data on the population structure of several widespread self-pollinating weeds, Allard (31) and Kannenberg and Allard (33) concluded that local populations could be composed of a very large number of different genotypes and that most individuals were heterozygous to some extent.

Today it is recognized that these viewpoints represent opposite ends of a spectrum of conditions found in inbreeding species (78). Furthermore, as Jain (78,79) has emphasized, it is also possible to find these extremes among populations of a single species. For example, populations of *Avena barbata* from California are largely monomorphic for morphologic and allozyme markers throughout much of the Central Valley, the Sierra Nevada foothills, and southern California. In contrast, in the adjacent coastal range, populations are polymorphic to a varying degree with highly variable populations occurring in the vicinity of the San Francisco Bay area (26,79). Hypotheses to account for these differences in the patterns of genetic variation include founder effect, small differences in outcrossing rates among populations, and perhaps introgressive hybridization with the cultivated oat *A. sativa* (79). Related studies in California on *A. fatua* have demonstrated substantial stores of genetic variation within populations such that no two individuals appear to be genetically similar (25,26,32).

In contrast, several recent surveys of enzyme variation in widespread inbreeding weed species have demonstrated almost no detectable genetic variation. The cattails *Typha latifolia* and *T. domingensis* are emergent perennial aquatics that are native and widespread throughout North America. Although they constitute important components of marshland vegetation, they are also common weeds of irrigation canals, drainage ditches, and cultivated rice fields (17). Mashburn et al. (80) sampled 74 populations of *T. latifolia* and 52 populations of *T. domingensis* from the southeastern United States. They found no variation in electrophoretic assays involving 10 enzyme systems. The study was recently extended to include more than 500 *Typha* stands from a broad geographic range in North America with similar findings. Sharitz and co-workers (80) suggest that the apparent lack of intraspecific allozyme variability in *Typha* may be associated with several factors, including extensive inbreeding in the species, genetic drift, and undetected variability at regulatory loci (80,81). The work is of particular interest because of earlier reports of ecotype formation in North American populations of *Typha* (82).

In Australia, Moran and Marshall (83) surveyed the degree of genetic polymorphism at 13 enzyme loci in 12 populations of the cocklebur (Noogoora burr) *Xanthium strumarium*. The species is a predominantly inbreeding, monoecious annual of worldwide distribution, reported as a weed in 11 crops from 28 countries (27). The species is highly polymorphic and has been divided into eight races on the basis of fruit morphology (84). Four of these races are naturalized in Australia, and all were included in Moran and Marshall's study. This study found very little variation within each of the four races, although they were genetically differentiated from one another. Within three of the races all individuals assayed were homozygous for the same alleles at all gene loci. Hence the races in Australia may be represented by a single genotype at these loci (83). It is not known whether founder effects are responsible for the allozyme uniformity of *X. strumarium* in Australia. Surveys of populations from the native New World range may resolve this question.

In a broad electrophoretic survey of 10 taxa in the New World *Solanum* section *Androceras*, Whalen (85) found that the eight annual weeds in the group displayed significantly lower levels of genetic variation than did the two perennial nonweeds. Of the 19 populations of the weed taxa included in the survey, most were genetically monomorphic at the seven gene loci assayed. Additional evidence of low levels of allozyme variation within and in some cases among populations of weed species come from the studies of Levin (86) on *Oenothera biennis*, Rick et al. (87) on *Lycopersicon pimpinellifolium*, Crawford and Wilson (88,89) on *Chenopodium* spp., and Moran on *Xanthium spinosum* and Marshall and Weiss on *Emex spinosa* (both unpublished, cited in Ref. 12).

In summary, although populations of inbreeding weeds can contain significant amounts of genetic variation, it appears that in a significant number of cases populations are markedly depauperate in allozyme variation when compared to nonweed species. This pattern could arise from genetic bottlenecks associated with colonization and as a result of habitual inbreeding. As is discussed later, some of these findings may have important implications for biological control.

Outcrossers. Outcrossing mechanisms such as self-incompatibility and dioecism restrict colonizing ability since population establishment requires the introduction of at least two compatible individuals to a site (73,74). Few widespread weeds exhibit such mechanisms, although exceptions include *Rumex* spp. and *Silene alba*, which are dioecious (54), and *Helianthus annuus* and *Turnera ulmifolia*, which possess self-incompatibility systems (90,91). Nevertheless, many self-compatible weeds are outbred to some extent, thus enhancing the buildup of genetic variation in populations.

There are relatively few studies of the levels of genetic variation in fully outbreeding weeds because of the general scarcity of such weeds. Among these, assays of enzyme polymorphisms in *Helianthus annuus* and *Lolium* spp. (cited in Ref. 4) have demonstrated substantial amounts of genetic variation both within and among populations. Also, surveys of a large number of nonweed species (4,39,41) indicate that the overall levels of genetic variation in outbreeding species are significantly higher than those of inbreeders (Table 4).

HYBRIDIZATION AND POLYPLOIDY

The migration of weeds into unoccupied territory can provide opportunities for hybridization with related taxa. In some cases one of the hybridizing species may be native to the area, as in the case of hybridization between the native *Silene dioica* and the introduced weed *S. alba* in Great Britain (92). Alternatively, several species may be sympatric in the alien range and produce hybrids, as in *Tragopogon* spp. in western North America (see following paragraphs and Ref. 93). In either case hybridization and the establishment of hybrid progeny may be favored because of the disturbed, open nature of many weed habitats. The products of

TABLE 4.   Levels of Electrophoretically Detectable Genetic Variation in Plants from Different Stages of Succession and with Varying Patterns of Mating

Plants from Different Stages of Succession and with Varying Mating Systems	Number of Species (N)	Mean No. of Loci	Percentage of Loci Polymorphic (P)	Number of Alleles per Locus (A)	Polymorphic Index[a] (PI)
Stages of Succession					
Weedy and early	54	12.5	29.7	1.6	0.116
Middle	49	9.7	37.9	1.6	0.137
Late	10	12.0	62.8	2.1	0.271
Mating System					
Selfed	33	14.2	17.9	1.3	0.058
Mixed	42	8.6	14.2	1.8	0.181
Outcrossed	36	11.3	51.1	1.9	0.185

Source: After Hamrick et al. (4). Reproduced, with permission, from the *Annual Review of Ecology and Systematics*, Vol. 10, copyright 1979 by Annual Reviews, Inc.

[a]$P < .01$ following ANOVA.

hybridization vary and can include (1) new more vigorous genotypes, as in *Carduus nutans* and *C. acanthoides* in Canada (94) and *Emex spinosa* and *E. australis* in Australia (95); (2) sterile, but vegetatively vigorous types such as *Salvinia molesta* (48,49); (3) hybrid swarms as in *Raphanus sativus* and *R. raphanistrum* (96) and *Helianthus* spp. (90,97,98) in California; and (4) formation of new species, such as *Spartina anglica* (99,100), *Galeopsis tetrahit* (101), and *Lamium moluccellifolium* (102) in Europe. Usually the resultant hybrid populations contain substantially more genetic variability than do those of their progenitors.

Several examples serve to illustrate the point. *Helianthus annuus* (common sunflower) is a widespread annual weed of ruderal habitats that is native to parts of North America. It is unusual among annual weeds in exhibiting several features not normally associated with colonizing ability. These features are self-incompatibility, a relatively low reproductive capacity, and weakly developed seed dispersal and dormancy mechanisms. However, the species is highly variable both morphologically and physiologically, and Heiser (90) has suggested that this variation may be the key to its success as a weed. Heiser's detailed studies of annual *Helianthus* spp. (90,103,104) suggest that the great variability of *H. annuus* has resulted from frequent hybridization with related species during its westward migration in association with humans. Enrichment of the gene pool of *H. annuus* may have enabled colonization of a broad range of habitats. *Helianthus annuus* may be considered a compilospecies, that is, "a genetically aggressive species, plundering related species of their heredities" (105). Much of the variability in other weed complexes may arise from similar processes, although it may often be difficult to distinguish the effects of introgressive hybridization from simple ecotypic differentiation (98,106).

Hybridization can set the stage for polyploid formation and the evolution of new species. This has occurred among biennial *Tragopogon* spp. (salsify or goat's beard), which are common ruderal weeds in North America and native to the Old World. The three diploid species, *T. dubius, T. porrifolius,* and *T. pratensis,* are frequently found together and sterile F_1 hybrids are formed (93). As a result of the natural doubling of the chromosomes of F_1 hybrids, two tetraploid species, *T. mirus* and *T. miscellus,* were formed during this century in populations from Washington and Idaho. Using morphologic and karyologic evidence, Ownbey (93) was able to document the parentage of these new weeds, which are the only unambiguous examples of the recent natural origin of allotetraploid plant species.

In an effort to verify the parentage of the new *Tragopogon* species and to compare the patterns of electrophoretic variation in the diploids and tetraploids, Roose and Gottlieb (107) analyzed variability in 13 enzyme systems coded by a minimum of 21 gene loci. The study confirmed the proposed phylogeny of Ownbey (93) and revealed striking differences in the patterns of genetic variation among the five species. Populations of the three diploids were monomorphic, or nearly so, for different alleles at about 40% of their genes. Only six of the 292 individuals sampled were heterozygous at a single gene. In contrast, the tetraploids exhibited a fixed heterozygous multienzyme, phenotype specified, respectively, by 43% (*T. mirus*) and 33% (*T. miscellus*) of the 21 duplicated genes examined. With few exceptions, heterozygosity was displayed by all tetraploid individuals. Roose and Gottlieb (107) suggested that substantial levels of fixed heterozygosity and the resultant enzyme multiplicity may contribute to the wide ecological amplitude of polyploid species. It is noteworthy that since its origin *T. miscellus* has become a common weed in the vicinity of Spokane, Washington.

Nearly all the world's most serious weeds are polyploids (12). Presumably, the hybrid origin of many polyploids and the resultant high levels of fixed heterozygosity confers greater tolerance to environmental heterogeneity (107,108). In addition to the intergenomic diversity per homoeologous set of genes, there is evidence that polyploids may be more variable intragenomically than diploids (12). Thus among weedy groups it is probable that taxa with high chromosome numbers produce a variety of recombinant progeny and maintain a greater degree of genetic variation in populations than do their related diploid species.

ENVIRONMENTAL HETEROGENEITY

Although the genetic system and in particular the pattern of mating may strongly influence the genetic composition of offspring, the selective exigencies of the environment ultimately mold the genetic structure of populations. Theoretical studies (109,110) and limited empirical evidence (111,112) suggest a positive relationship between the levels of genetic variation in populations and the degree of environmental heterogeneity. Apart from the problem of measuring and comparing environmental heterogeneity among habitats, it is often difficult to determine the direction of causality. Do heterogeneous environments select for genetic diversity, or are populations with high genetic variation fitter in variable environments?

In the survey of Hamrick et al. (4), taxa from weedy (disturbed) and early successional habitats were significantly less variable than those from the middle and climax stages of succession (Table 4). Because of their simple structure and relatively low biotic diversity, weed habitats are generally considered less heterogeneous than most other vegetation types (4,113). Considering just weed habitats and assuming that a relationship does exist between genetic variation and environmental heterogeneity, we may make several predictions. First, populations of agrestals should in general exhibit lower genetic variation than those of ruderals or range-land weeds. This follows from differences between the habitats occupied by these weed groups. Agroecosystems, particularly those involving monocultures of annual crops, are deliberately made uniform by the agri-culturalist and can be highly predictable in time and space. In contrast, ruderal habitats tend to be more complex, and in grassland habitats con-siderable biotic and edaphic diversity may occur. Whereas weeds of crops may be surrounded by a single genetic strain, ruderal and pasture weeds may interact with a variety of neighbors. Second, we may expect that specialized crop weeds such as crop mimics that are restricted to a par-ticular crop (e.g., *Echinochloa crus-galli* var. *oryzicola* in rice) may be less variable than agrestals capable of infesting a wide range of different crops (e.g., *E. crus-galli* var. *crus-galli*, *Cynodon dactylon*, and *Cyperus rotundus*). Unfortunately, most electrophoretic studies of genetic varia-tion in weeds have so far dealt with widespread ruderal species. Surveys of genetic variation in agricultural weeds are also required.

IMPLICATIONS FOR BIOLOGICAL CONTROL

Diseases and insects exert strong selective pressures on the genetic systems and evolution of plant populations (113,114). The greatest forces are ex-erted on long-lived, relatively stable communities, especially tropical climax communities. The weakest pathogen and herbivore pressures ostensibly occur in transient, weedy communities of the temperate regions. Levin (113) has suggested that this pattern is a function of host predictability. In a North American survey, annual plants (mostly weeds) were found to have fewer pathogens per unit area or vegetational zone than did perennial herbs, shrubs, or trees (113). Nevertheless, weeds are invariably attacked by diseases and insects, and there has been a growing recognition over the past two decades that plant pathogens have a poten-tial for the control of weeds (116–118).

An important issue concerns the likelihood of weeds evolving disease resistance after release of a pathogen into an area or during repeated ap-plications of a microbial herbicide. Because of our general ignorance of the processes of coevolution in natural communities, it is difficult to make precise predictions of the outcome of such weed-pathogen interactions. Much will depend on the levels and nature of genetic variation in both the host and the pathogen. Two possibilities could arise. In the first case, resistance genes may be absent from the gene pool of the weed in the introduced region. Whether resistance would arise and how long this might take are matters of speculation. Nelson (119) has suggested that in natural communities the evolution of *de novo* resistance to a "new" parasite is probably accomplished by a simple genetic change at a single locus resulting in "race-specific" resistance. Subsequent coevolution would result in the ultimate accumulation of many resistance and fitness genes in the host and pathogen, and the establishment of a host-pathogen genetic equilibrium (120). Alternatively, resistance genes may already be present in the gene pool of the weed as a result of previous coevolutionary interactions with the pathogen in other regions of the world. It is uncertain whether such genes would be maintained in the absence of the specific plant pathogen. It is possible that they would be lost either because there is some cost associated with their maintenance or through random mutations. Leonard and Czochor (115) have recently reviewed much of the theoretical work concerned with the population genetics and evolution of host-pathogen interactions, but more empirical studies are required before firm conclusions can be reached (see also Chapter 7).

It is relevant at this point to consider the responses of weed populations to herbicides. Despite the extensive and repeated use of chemical control methods during the past 30 years, relatively few cases of the evolution of genetic resistance to specific herbicides have been reported, although there are numerous examples of variation in response of weed populations to herbicides (121). The only cases where genetic resistance has developed in the field as a result of herbicide treatments involve the S-triazines, atrazine and simazine. All taxa involved are temperate annuals and include *Ambrosia artemisiifolia* (122), *Amaranthus retroflexus* (123), *Brassica campestris* (124), *Capsella bursa-pastoris* (125), *Chenopodium album* (126), *Chenopodium strictum* (127), *Poa annua* (128), and *Senecio vulgaris* (129).

Gressel and Segel (130) have explored some of the possible reasons for the limited number of cases of the evolution of genetic resistance to herbicides in weed species. Several of the factors they have discussed are relevant to the microbial herbicide situation. Of particular importance are the selection pressures exerted by the control agent, the degree of phenotypic plasticity exhibited by the weed, and the dormancy and germination behavior of the weed seed reservoir. For each factor, the important consideration is the presence in the weed population of significant numbers of surviving, susceptible genotypes. In the case of the weed seed reservoir, the buffering action of the seed bank of dormant, susceptible genotypes is likely to retard the development of resistance, particularly in weed species capable of some outcrossing.

In future microbial herbicide research it will be particularly important to determine whether weed survival results from true genetic resistance or simply from environmental factors (microclimate, application efficiency, etc.). Progeny tests of surviving individuals, followed by exposure to the plant pathogen under test, should provide an answer to this question. It is probable that in many cases an interaction of genetic and environmental factors will influence the degree of susceptibility exhibited by the host.

It has been suggested that species with reproductive systems that restrict recombination, and hence favor genetic uniformity, are more likely to succumb to disease outbreaks (57,113,131,132). In this regard, it is of interest that many of the successful cases of biological control of weeds have involved primarily asexually reproducing species (e.g., apomicts such as *Chondrilla juncea, Eupatorium adenophorum, Hypericum perforatum, Opuntia* spp., and *Rubus penetrans*). Burdon and Marshall (133) examined the degree of biological control achieved in 81 different attempts in a total of 45 weed species. They demonstrated a significant correlation between the degree of control and the predominant mode of reproduction of the target species. Asexually reproducing species were effectively controlled more often than those reproducing by sexual means (Table 5). Presumably, successful biological control is favored by limited amounts of genetic variation in weed populations. If this is true, many aquatic weeds that reproduce predominantly by clonal propagation (e.g., *Alternanthera philoxeroides, Elodea canadensis, Eichhornia crassipes, Hydrilla verticillata, Pistia stratiotes,* and *Salvinia molesta*) would seem to be excellent targets for biological control. The low levels or absence of sexual reproduction in many populations of these species would reduce the likelihood of coevolutionary responses following the introduction of a plant pathogen.

TABLE 5. Reproductive System and Average Degree of Control of Weeds Utilized in Biological Control Program

Major Reproductive Mode	Degree of Control Achieved (%)			
	None	Partial	Substantial	Complete
Asexual (n = 24 spp.)	4.2	8.1	39.0	48.7
Sexual (n = 16 spp.)	18.7	56.7	15.6	8.9

Source: Burdon and Marshall (133).

A final word of caution is required in this discussion of the relationship between genetic variation and biological control. It is quite possible that in many weed species the actual amount of genetic variation in populations will be a poor predictor of whether successful control can be achieved. What is probably more important is the nature of the genetic variation present in populations and how this will influence the rate at which host resistance may evolve under field conditions. Unfortunately, at this time little is known of the evolution and maintenance of host resistance in natural plant populations.

CONCLUSIONS

Compared with other plant life forms, populations of weed species frequently exhibit limited genetic variation. Such limited genetic variation may be the result of genetic bottlenecks associated with repeated episodes of colonization, extensive clonal propagation, inbreeding, and the relative environmental homogeneity of many agroecosystems. Nevertheless, populations of some weed species contain substantial stores of genetic variation, and interpopulation genetic differentiation is a pervasive feature of widely distributed weeds. The development of a high level of genetic diversity among weed populations is fostered by multiple seed introductions, habitat longevity, environmental heterogeneity, large population size, outbreeding, and hybridization with related taxa. Limited evidence involving biological control agents suggest that asexually reproducing weeds are more easily controlled than sexual species. This difference may be associated with the lower levels of genetic variation present in weed species with restricted recombination systems.

REFERENCES

1. R. A. Fisher. 1958. *The Genetical Theory of Natural Selection*, 2nd rev. ed. Dover Publications, New York, 291 pp.

2. R. K. Selander. 1976. Genic variation in natural populations. Pp. 21–45 in: F. J. Ayala, ed., *Molecular Evolution*. Sinauer Associates, Sunderland, MA.

3. E. Nevo. 1978. Genetic variation in natural populations: Patterns and theory. *Theor. Popul. Biol.* 13, 121–177.

4. J. L. Hamrick, Y. B. Linhart, and J. B. Mitton. 1979. Relationships between life history characteristics and electrophoretically detectable genetic variation in plants. *Annu. Rev. Ecol. Syst.* 10, 173–200.

5. H. G. Baker. 1974. The evolution of weeds. *Annu. Rev. Ecol. Syst.* 5, 1–24.

6. A. H. Bunting. 1960. Some reflections on the ecology of weeds. Pp. 11–26 in: J. L. Harper, ed., *The Biology of Weeds*. Blackwell, Oxford, England.

7. L. J. King. 1966. *Weeds of the World: Biology and Control.* Hill, London, 526 pp.

8. J. R. Harlan. 1975. *Crops and Man.* American Society of Agronomy and Crop Science Society of America, Madison, WI, 295 pp.

9. H. G. Baker. 1965. Characteristics and modes of origin of weeds. Pp. 147–172 in: H. G. Baker and G. L. Stebbins, eds., *The Genetics of Colonizing Species.* Academic Press, New York.

10. S. K. Jain and P. S. Martins. 1979. Ecological genetics of the colonizing ability of rose clover (*Trifolium hirtum* All.). *Am. J. Bot.* **66**, 361–366.

11. S. C. H. Barrett and B. F. Wilson. 1981. Colonizing ability in the *Echinochloa crus-galli* complex (barnyard grass). I. Variation in life history. *Can. J. Bot,* **59**, 1844–1860.

12. A. H. D. Brown and D. R. Marshall. 1981. Evolutionary changes accompanying colonization in plants. Pp. 351–353 in: G. G. E. Scudder and J. L. Reveal, eds., *Proceedings of the Second International Congress of Systematic and Evolutionary Biology.* Hunt Institute for Botanical Documentation, Carnegie-Mellon University, Pittsburg, PA.

13. H. Godwin. 1960. The history of weeds in Britain. Pp. 1–10 in: J. L. Harper, ed., *The Biology of Weeds.* Blackwell, Oxford, England.

14. H. G. Baker. 1972. Migrations of weeds. Pp. 327–347 in: D. H. Valentine, ed., *Taxonomy, Phytogeography and Evolution.* Academic Press, London.

15. J. McNeill. 1976. The taxonomy and evolution of weeds. *Weed Res.* **16**, 399–413.

16. A. Gray. 1879. The pertinacity and predominance of weeds. *Am. J. Sci.* **118**, 161–167.

17. S. C. H. Barrett and D. E. Seaman. 1980. The weed flora of Californian rice fields. *Aquat. Bot.* **9**, 351–376.

18. C. B. Huffaker. 1971. *Biological Control.* Plenum Press, New York, 511 pp.

19. R. A. Kennedy, S. C. H. Barrett, D. Vander Zee, and M. E. Rumpho. 1980. Germination and seedling growth under anaerobic conditions in *Echinochloa crus-galli* (barnyard grass). *Plant Cell Environ.* **3**, 243–249.

20. R. C. Lewontin. 1974. *The Genetic Basis of Evolutionary Change.* Columbia University Press, New York, 346 pp.

21. J. L. Harper and D. Gajic. 1961. Experimental studies on the mortality and plasticity of a weed. *Weed Res.* **1**, 91–104.

22. A. D. Bradshaw. 1965. Evolutionary significance of phenotypic plasticity in plants. *Adv. Genet.* **13**, 115–155.

23. D. R. Marshall and S. K. Jain. 1968. Phenotypic plasticity of *Avena fatua* and *A. barbata*. *Am. Nat.* **102**, 457–467.

24. T. N. Cooley and D. F. Martin. 1978. Seeking "super hyacinths." *J. Environ. Sci. Health* **13**, 469–479.

25. S. K. Jain and D. R. Marshall. 1967. Population studies in predominantly self-pollinating species. X. Variation in natural populations of *Avena fatua* and *A. barbata*. *Am. Nat.* **101**, 19–33.

26. S. K. Jain. 1969. Comparative ecogenetics of two *Avena* species occurring in central California. *Evol. Biol.* **3**, 73–118.

27. L. G. Holm, D. L. Plucknett, J. V. Pancho, and H. P. Herberger. 1977. *The World's Worst Weeds: Distribution and Biology.* The University Press of Hawaii, Honolulu, HI, 609 pp.

28. P. H. Davis and V. H. Heywood. 1963. *Principles of Angiosperm Taxonomy.* Van Nostrand, Princeton, NJ, 556 pp.

29. J. Heslop-Harrison. 1964. Forty years of genecology. *Adv. Ecol. Res.* **2**, 159–247.

30. D. R. Marshall and A. H. D. Brown. 1975. Optimum sampling strategies in genetic conservation. Pp. 53–80 in: O. H. Frankel and J. G. Hawkes, eds., *Crop Genetic Resources for Today and Tomorrow.* Cambridge University Press, Cambridge, England.

31. R. W. Allard. 1965. Genetic systems associated with colonizing ability in predominantly self-pollinated species. Pp. 49–76 in: H. G. Baker and G. L. Stebbins, eds., *The Genetics of Colonizing Species.* Academic Press, New York.

32. A. G. Imam and R. W. Allard. 1965. Population studies in predominantly self-pollinated species. VI. Genetic variability between and within natural populations of wild oats from differing habitats in California. *Genetics* **51**, 49–62.

33. L. W. Kannenberg and R. W. Allard. 1967. Population studies in predominantly self-pollinated species. VIII. Genetic variability in the *Festuca microstachys* complex. *Evolution* **21**, 227–240.

34. S. K. Jain, D. R. Marshall, and K. Wu. 1970. Genetic variability in natural populations of softchess (*Bromus mollis* L.). *Evolution* **24**, 649–659.

35. J. L. Hamrick and R. W. Allard. 1975. Correlations between quantitative characters and enzyme genotypes in *Avena barbata. Evolution* **29**, 438–442.

36. R. W. Allard. 1960. *Principles of Plant Breeding.* Wiley, New York, 485 pp.

37. R. C. Lewontin and J. L. Hubby. 1966. A molecular approach to the study of genic heterozygosity in natural populations. II. Amount of variation and degree of heterozygosity in natural populations of *Drosophila pseudoobscura. Genetics* **54**, 595–609.

38. F. J. Ayala, ed. 1976. *Molecular Evolution.* Sinauer Associates, Sunderland, MA, 277 pp.

39. A. H. D. Brown. 1979. Enzyme polymorphism in plant populations. *Theor. Popul. Biol.* **15**, 1–42.

40. R. S. Singh, R. C. Lewontin, and A. A. Felton. 1976. Genetic heterogeneity within electrophoretic "alleles" of xanthine dehydrogenase in *Drosophila pseudoobscura. Genetics* **84**, 609–629.

41. L. D. Gottlieb. 1977. Electrophoretic evidence and plant systematics. *Ann. Mo. Bot. Gard.* **64**, 161–180.

42. L. D. Gottlieb. 1976. Biochemical consequences of speciation in plants. Pp. 123–140 in: F. J. Ayala, ed., *Molecular Evolution.* Sinauer Associates, Sunderland, MA.

43. J. L. Hamrick. 1979. Genetic variation and longevity. Pp. 84–113 in: O. T. Solbrig, S. Jain, G. B. Johnson, and P. H. Raven, eds., *Topics in Plant Population Biology.* Columbia University Press, New York.

44. G. A. Mulligan. 1965. Recent colonization by herbaceous plants in Canada. Pp. 127–146 in: H. G. Baker and G. L. Stebbins, eds., *The Genetics of Colonizing Species.* Academic Press, New York.

45. H. Wild. 1968. *Weeds and Aliens in Africa: The American Immigrant.* University College of Rhodesia, Salisbury, Rhodesia, 30 pp.

46. E. Mayr. 1963. *Animal Species and Evolution.* Belknap Press of Harvard University Press, Cambridge, MA, 797 pp.

47. C. D. Sculthorpe. 1967. *The Biology of Aquatic Vascular Plants.* Edward Arnold, London, 610 pp.

48. C. D. K. Cook and B. J. Gut. 1971. *Salvinia* in the state of Kerala, India. *PANS* **17**, 438–447.

49. D. S. Mitchell. 1972. The Kariba weed: *Salvinia molesta. Br. Fern Gaz.* **10**, 251–252.

50. R. D. Blackburn, L. W. Weldon, R. R. Yeo, and T. M. Taylor. 1969. Identification and distribution of certain similar-appearing submersed aquatic weeds in Florida. *Hyacinth Control J.* **8**, 17–21.

51. S. C. H. Barrett. 1977. Tristyly in *Eichhornia crassipes* (Mart.) Solms (water hyacinth). *Biotropica* **9**, 230–238.

52. S. C. H. Barrett. 1980. Sexual reproduction in *Eichhornia crassipes* (water hyacinth). I. Fertility of clones from diverse regions. *J. Appl. Ecol.* **17**, 101–112.

53. S. C. H. Barrett. 1980. Sexual reproduction in *Eichhornia crassipes* (water hyacinth). II. Seed production in natural populations. *J. Appl. Ecol.* **17**, 113–124.

54. P. A. Fryxell. 1957. Mode of reproduction of higher plants. *Bot. Rev.* **23**, 135–233.

55. H. G. Baker. 1959. Reproductive methods as factors in speciation in flowering plants. *Cold Spring Harbor Symp. Quant. Biol.* **24**, 177–191.

56. R. Frankel and E. Galun. 1977. *Pollination Mechanisms, Reproduction, and Plant Breeding.* Springer-Verlag, Berlin, Federal Republic of Germany, 281 pp.

57. J. L. Harper. 1977. *Population Biology of Plants.* Academic Press, London, 892 pp.

58. R. J. Moore and D. R. Lindsay. 1953. Fertility and polyploidy in *Euphorbia cyparissias* in Canada. *Can. J. Bot.* **31**, 152–163.

59. T. Pritchard. 1960. Race formation in weedy species with special reference to *Euphorbia cyparissias* L. and *Hypericum perforatum* L. Pp. 61–66 in: J. L. Harper, ed., *The Biology of Weeds.* Blackwell, Oxford, England.

60. M. G. Cahn and J. L. Harper. 1976. The biology of the leaf mark polymorphism in *Trifolium repens* L. I. Distribution of phenotypes at a local scale. *Heredity* **37**, 309–325.

61. J. A. Silander. 1979. Microevolution and clone structure in *Spartina patens. Science* **203**, 658–660.

62. I. D. Soane and A. R. Watkinson. 1979. Clonal variation in populations of *Ranunculus repens. New Phytol.* **82**, 557–573.

63. A. Nygren. 1954. Apomixis in the angiosperms. *Bot. Rev.* **20**, 577–649.

64. V. J. Hull and R. H. Groves. 1973. Variation in *Chondrilla juncea* L. in south-eastern Australia. *Aust. J. Bot.* **21**, 113–135.

65. M. Costas-Lippmann. 1979. Embryogeny of *Cortaderia selloana* and *C. jubata* (Gramineae). *Bot. Gaz.* **140**, 393–397.

66. A. J. Richards. 1970. Eutriploid facultative agamospermy in *Taraxacum. New Phytol.* **69**, 761–774.

67. J. Clausen. 1954. Partial apomixis as an equilibrium system in evolution. *Caryologia* **6**, (suppl.) 469–479.

68. O. T. Solbrig and B. B. Simpson. 1974. Components of regulation of a population of dandelions in Michigan. *J. Ecol.* **62**, 473–486.

69. O. T. Solbrig and B. B. Simpson. 1977. A garden experiment on competition between biotypes of the common dandelion (*Taraxacum officinale*). *J. Ecol.* **65**, 427–430.

70. J. A. Usberti, Jr. and S. K. Jain. 1978. Variation in *Panicum maximum*: A comparison of sexual and asexual populations. *Bot. Gaz.* **139**, 112–116.

71. J. Cardenas, C. E. Reyes, and J. D. Doll. 1972. *Tropical Weeds.* Vol. 1. Instituto Colombiano Agropêcuario, Bogotá, Columbia, 341 pp.

72. R. B. Knox and J. Heslop-Harrison. 1963. Experimental control of aposporous apomixis in a grass of the Andropogoneae. *Bot. Not.* **116**, 127–141.

73. H. G. Baker. 1955. Self-compatibility and establishment after "long-distance" dispersal. *Evolution* 9, 347–349.

74. G. L. Stebbins. 1957. Self fertilization and population variability in the higher plants. *Am. Nat.* 91, 337–354.

75. J. Antonovics. 1968. Evolution in closely adjacent plant populations. V. Evolution of self-fertility. *Heredity* 23, 219–238.

76. C. D. Darlington and K. Mather. 1950. *The Elements of Genetics*. Macmillan, New York, 446 pp.

77. R. W. Allard, S. K. Jain, and P. L. Workman. 1968. The genetics of inbreeding populations. *Adv. Genet.* 14, 55–131.

78. S. K. Jain. 1976. Evolution of inbreeding in plants. *Annu. Rev. Ecol. Syst.* 7, 469–495.

79. S. K. Jain. 1975. Population structure and the effects of breeding system. Pp. 15–36 in: O. H. Frankel and J. G. Hawkes. eds., *Crop Genetic Resources for Today and Tomorrow*. Cambridge University Press, Cambridge, England.

80. S. J. Mashburn, R. R. Sharitz, and M. H. Smith. 1978. Genetic variation among *Typha* populations of the southeastern United States. *Evolution* 32, 681–685.

81. K. Krattinger. 1975. Genetic mobility in *Typha*. *Aquat. Bot.* 1, 57–70.

82. S. J. McNaughton. 1966. Ecotype function in the *Typha* community-type. *Ecol. Monogr.* 36, 297–325.

83. G. F. Moran and D. R. Marshall. 1978. Allozyme uniformity within and variation between races of the colonizing species *Xanthium strumarium* L. (Noogoora burr). *Aust. J. Biol. Sci.* 31, 283–291.

84. D. Löve and P. Dansereau. 1959. Biosystematic studies on *Xanthium*: Taxonomic appraisal and ecological status. *Can. J. Bot.* 37, 173–208.

85. M. D. Whalen. 1979. Allozyme variation and evolution in *Solanum* section *Androceras*. *Syst. Bot.* 4, 203–222.

86. D. A. Levin. 1975. Genic heterozygosity and protein polymorphism among local populations of *Oenothera biennis*. *Genetics* 79, 477–491.

87. C. M. Rick, J. F. Fobes, and M. Holle. 1977. Genetic variation in *Lycopersicon pimpinellifolium*: Evidence of evolutionary change in mating systems. *Plant Syst. Evol.* 127, 139–170.

88. D. J. Crawford and H. D. Wilson. 1977. Allozyme variation in *Chenopodium fremontii*. *Syst. Bot.* 2, 180–190.

89. D. J. Crawford and H. D. Wilson. 1979. Allozyme variation in several closely related diploid species of *Chenopodium* of the western United States. *Am. J. Bot.* 66, 237–244.

90. C. B. Heiser, Jr. 1965. Sunflowers, weeds, and cultivated plants. Pp. 391–401 in: H. G. Baker and G. L. Stebbins, eds., *The Genetics of Colonizing Species*. Academic Press, New York.

91. S. C. H. Barrett. 1978. Heterostyly in a tropical weed: The reproductive biology of the *Turnera ulmifolia* complex (Turneraceae). *Can. J. Bot.* 56, 1713–1725.

92. H. G. Baker. 1948. Stages in invasion and replacement demonstrated by species of *Melandrium*. *J. Ecol.* 36, 96–119.

93. M. Ownbey. 1950. Natural hybridization and amphiploidy in the genus *Tragopogon*. *Am. J. Bot.* 37, 487–499.

94. G. A. Mulligan and R. J. Moore. 1961. Natural selection among hybrids between *Carduus acanthoides* and *C. nutans* in Ontario. *Can. J. Bot.* 39, 269–279.

95. E. Putievsky, P. W. Weiss, and D. R. Marshall. 1980. Interspecific hybridization between *Emex australis* and *E. spinosa*. *Aust. J. Bot.* **28**, 323-328.

96. C. Panetsos and H. G. Baker. 1968. The origin of variation in "wild" *Raphanus sativus* (Cruciferae) in California. *Genetica* **38**, 243-274.

97. G. L. Stebbins and K. Daly. 1961. Changes in the variation pattern of a hybrid population of *Helianthus* over an eight-year period. *Evolution* **15**, 60-71.

98. A. M. Olivieri and S. K. Jain. 1977. Variation in the *Helianthus exilis-bolanderi* complex: A reexamination. *Madroño* **24**, 177-189.

99. C. J. Marchant. 1967. Evolution in *Spartina* (Gramineae). I. The history and morphology of the genus in Britain. *J. Linn. Soc. (Bot.)* **60**, 1-24.

100. C. J. Marchant. 1968. Evolution in Spartina (Gramineae). II. Chromosomes, basic relationships and the problem of S. X *Townsendii* agg. *J. Linn. Soc. (Bot.)* **60**, 381-409.

101. A. Müntzing. 1930. Outlines to a genetic monograph of the genus *Galeopsis* with special reference to the nature and inheritance of partial sterility. *Hereditas* **13**, 185-341.

102. P. Bernström. 1955. Cytogenetic studies on relationships between annual species of *Lamium*. *Hereditas* **41**, 1-122.

103. C. B. Heiser, Jr. 1949. Study in the evolution of the sunflower species *Helianthus annuus* and *H. bolanderi*. *Univ. Calif. Publ. Bot.* **23**, 157-208.

104. C. B. Heiser, Jr. 1954. Variation and subspeciation in the common sunflower, *Helianthus annuus*. *Am. Midl. Nat.* **51**, 287-305.

105. J. R. Harlan and J. M. J. de Wet. 1963. The compilospecies concept. *Evolution* **17**, 497-501.

106. C. B. Heiser, Jr. 1973. Introgression reexamined. *Bot. Rev.* **39**, 347-366.

107. M. L. Roose and L. D. Gottlieb. 1976. Genetic and biochemical consequences of polyploidy in *Tragopogon*. *Evolution* **30**, 818-830.

108. G. R. Babbel and R. P. Wain. 1977. Genetic structure of *Hordeum jubatum*. I. Outcrossing rates and heterozygosity levels. *Can. J. Genet. Cytol.* **19**, 143-152.

109. J. Antonovics. 1971. The effects of a heterogeneous environment on the genetics of natural populations. *Am. Sci.* **59**, 593-599.

110. P. W. Hedrick, M. E. Ginevan, and E. P. Ewing. 1976. Genetic polymorphism in heterogeneous environments. *Annu. Rev. Ecol. Syst.* **7**, 1-32.

111. J. F. McDonald and F. J. Ayala. 1974. Genetic response to environmental heterogeneity. *Nature* **250**, 572-574.

112. R. W. Allard, R. D. Miller, and A. L. Kahler. 1978. The relationship between degree of environmental heterogeneity and genetic polymorphism. Pp. 49-69 in: A. H. J. Freyson and J. W. Woldendorp, eds., *Structure and Functioning of Plant Populations*. Elsevier North-Holland, Amsterdam, The Netherlands.

113. D. A. Levin. 1975. Pest pressure and recombination systems in plants. *Am. Nat.* **109**, 437-451.

114. D. H. Janzen. 1968. Host plants as islands in evolutionary and contemporary time. *Am. Nat.* **102**, 592-595.

115. K. J. Leonard and R. J. Czochor. 1980. Theory of genetic interactions among populations of plants and their pathogens. *Annu. Rev. Phytopathol.* **18**, 237-258.

116. C. L. Wilson. 1969. Use of plant pathogens in weed control. *Annu. Rev. Phytopathol.* **7**, 411–434.

117. F. W. Zettler and T. E. Freeman. 1972. Plant pathogens as biocontrols of aquatic weeds. *Annu. Rev. Phytopathol.* **10**, 455–470.

118. G. E. Templeton, D. O. TeBeest, and R. J. Smith, Jr. 1979. Biological weed control with mycoherbicides. *Annu. Rev. Phytopathol.* **17**, 301–310.

119. R. R. Nelson. 1978. Genetics of horizontal resistance to plant diseases. *Annu. Rev. Phytopathol.* **16**, 359–378.

120. R. R. Nelson. 1979. Some thoughts on the coevolution of plant pathogenic fungi and their hosts. Pp. 17–25 in: B. B. Nickol, ed., *Host-Parasite Interfaces*. Academic Press, New York.

121. J. L. Hammerton. 1967. Intra-specific variations in susceptibility to herbicides. *Meded. Rijksfac. Landbouwwet. Gent.* **32**, 999–1012.

122. V. Souza Machado, J. D. Bandeen, W. D. Taylor, and P. Lavigne. 1977. Atrazine resistant biotypes of common ragweed and birds rape. *Res. Rep. Can. Weed Comm. (East. Sec.)* **22**, 305.

123. S. R. Radosevich. 1977. Mechanism of atrazine resistance in lambsquarters and pigweed. *Weed Sci.* **25**, 316–318.

124. B. Maltais and C. J. Bouchard. 1978. Une moutarde des oiseaux (*Brassica rapa* L.) résistante à l'atrazine. *Phytoprotection* **59**, 117–119.

125. R. J. Holliday and P. D. Putwain. 1974. Variation in susceptibility to simazine in three species of annual weeds. *Proc. Br. Weed Control Conf.* **12**, 649–654.

126. J. D. Bandeen and R. D. McLaren. 1976. Resistance of *Chenopodium album* to triazine herbicides. *Can. J. Plant Sci.* **56**, 411–412.

127. S. I. Warwick, V. Souza Machado, P. B. Marriage, and J. D. Bandeen. 1979. Resistance of *Chenopodium strictum* Roth (late-flowering goosefoot) to atrazine. *Can. J. Plant Sci.* **59**, 269–270.

128. J. M. Ducruet and J. Gasquez. 1978. Observation de la fluorescence sur feuille entière et mise en évidence de la résistance chloroplastique à l'atrazine chez *Chenopodium album* L. et *Poa annua* L. *Chemosphere* **8**, 691–696.

129. R. J. Holliday and P. D. Putwain. 1977. Evolution of resistance to simazine in *Senecio vulgaris* L. *Weed Res.* **17**, 291–296.

130. J. Gressel and L. A. Segel. 1978. The paucity of plants evolving genetic resistance to herbicides: Possible reasons and implications. *J. Theor. Biol.* **75**, 349–371.

131. M. W. Adams, A. H. Ellingboe, and E. C. Rossman. 1971. Biological uniformity and disease epidemics. *BioScience* **21**, 1067–1070.

132. J. J. Burdon, D. R. Marshall, and R. H. Groves. 1980. Aspects of weed biology important to biological control. Pp. 21–29 in: E. S. Del Fosse, ed., *Proceedings of the Fifth International Symposium on Biological Control of Weeds*. CSIRO, Canberra, Australia.

133. J. J. Burdon and D. R. Marshall. 1981. Biological control and the reproductive mode of weeds. *J. Appl. Ecol.* **18**, 649–658.

7

THE BENEFITS AND POTENTIAL HAZARDS OF GENETIC HETEROGENEITY IN PLANT PATHOGENS

K. J. LEONARD

U.S. Department of Agriculture, Science and Education Administration, Agricultural Research, Department of Plant Pathology, North Carolina State University, Raleigh, North Carolina

Genetic variation is a characteristic of all living organisms. Plant pathogens vary in virulence to their hosts, and host plants vary in resistance to their pathogens. We have seen enough epidemics on cultivated plants to appreciate the importance of genetic heterogeneity of hosts and pathogens. The genetic heterogeneity in the host serves as the basis for breeding disease-resistant cultivars; the heterogeneity in the pathogen provides the reservoir from which new virulent races may arise. When we consider the use of plant pathogens to control weeds, we should view the consequences of genetic heterogeneity in a new way. In many respects, our concerns in biological weed control are simply the reverse of those of the plant pathologist charged with protecting crops from epidemics, but there is an important difference. As Williams (1) put it, "The critical phase of biological control work against weeds is the selection of species that will not harm other plants, or at least useful plants. All other

Paper No. 6764 of the Journal Series of the North Carolina Agricultural Research Service, Raleigh.

considerations are subordinate. . . ." Following Williams' maxim, let us first consider the potential hazards of genetic heterogeneity in plant pathogens used as biological control agents.

Plant pathogens may be used against weeds in two principal ways: (1) as classical biocontrol agents, in which case pathogens found to be highly virulent to a weed are imported from another part of the world and are released to multiply, produce epidemics, and ultimately reduce the weed species to an inconsequential part of the ecosystem and (2) as microbial herbicides, which are applied as a massive dose of inoculum to the weeds at an early stage in each cropping season. Since they are not required to multiply or spread greatly, microbial herbicides may employ indigenous pathogens that do not ordinarily devastate their hosts under natural disease conditions (see Chapter 3 and Refs. 2–4).

This review is confined to plant pathogenic fungi because much more is known of the genetics of virulence in fungi than in nematodes, bacteria, mycoplasmas, or viruses. Furthermore, among all types of pathogens, fungi are most frequently considered as potential biocontrol agents.

HAZARDS OF GENETIC HETEROGENEITY—CLASSICAL BIOCONTROL

Importation of classical biocontrol agents to control weeds is a reversal of the principle of disease control in cultivated crops by quarantine. The dangers of foreign pest to crops and native plant species are well known. Before any plant pathogen is imported for release in a country where it did not occur previously, it must be subjected to rigorous screening to ensure that it will not injure beneficial plants in that country (see Chapter 11). According to Harris (5), it costs at least one scientist year (SY) to obtain a biocontrol agent and demonstrate that it is safe for release in Canada, and several additional SYs are required if the biocontrol agent is released.

As an example of the precautions normally taken, we may consider the case of *Puccinia chondrillina*, a rust fungus imported from Italy into Australia to control skeletonweed (*Chondrilla juncea*), an important pest of wheat fields and grasslands. The first isolates of the fungus were collected from France. Although they were highly virulent on French populations of skeletonweed, the isolates were discarded because they were avirulent on the Australian forms of the weed. This problem, evidently due to genetic heterogeneity in the weed and the pathogen, was overcome by more extensive collections in the Mediterranean area until the highly virulent Italian isolate was found (6). Once the strain of *P. chondrillina* that was most pathogenic to the Australian forms of skeletonweed had been discovered, its pathogenicity to 48 species of cultivated plants and 11 species of weeds related to *Chondrilla* was tested. *[and no wild plants?]* None of these species showed any symptoms (6). The fungus was released in Australia in 1971 and has spread rapidly from the release sites (7). There are no reports of any adverse effects of *P. chondrillina* on plants other than skeletonweed.

The isolate of *P. chondrillina* released in Australia successfully attacks the most common form of skeletonweed there, but it is avirulent on the two less common forms of the weed. The original isolate from Italy was moderately virulent to the two less common Australian forms of skeletonweed, but apparently it lost that virulence during the prolonged culture on the common form before its release in Australia (8). Thus the pathogenic potential of *P. chondrillina* is not fully represented by the introduced strain. If additional strains of the pathogen are introduced into Australia, they will also need to be screened for virulence to useful plants. Therefore, it might have been more efficient to screen the cultivated, nontarget plant species and related weeds against a mixture of the European collections of *P. chondrillina* rather than against the single strain selected for importation. The mixture of strains would have given a more complete representation of the pathogenic potential of the fungus.

The strain of *P. chondrillina* imported into Australia might eventually extend its host range to include all three Australian forms of skeletonweed. The virulence it lost during culture, apparently by mutation, might be regained by reverse mutation. Spontaneous mutations from avirulence to virulence in rust fungi have been demonstrated by Watson (9) for *Puccinia graminis*, by Zimmer et al. (10) for *P. coronata*, and by Flor (11) for *Melampsora lini*. These mutations served to extend the pathogen's host range to additional cultivars within the host species. However, it is unlikely that simple mutations would be sufficient to extend the host range of *P. chondrillina* to species other than skeletonweed. Crosses between *formae speciales* of *P. graminis* have shown that the inheritance of specificity at the host species level is complex and that the hybrids usually have much reduced virulence to the hosts of one or both parents (12,13). On the basis of this knowledge, we may conclude that simple mutations in rust fungi are unlikely to extend their host ranges to new species. Still, in evaluating the host range of *P. chondrillina*, it would have been safer to expose the potential host plants to a heterogeneous population of the fungus rather than to a single isolate.

With plant pathogens that are not obligate parasites, the adaptation to a new host species may occur more readily than it does with rust fungi. For example, isolates of *Phytophthora infestans* from potato are usually very weakly virulent on tomato. Although lesions on lower leaves of tomato plants may occasionally enlarge, the fungus usually induces only very small lesions, and in these lesions the fungus dies within 4 or 5 days (14). No sporulation occurs on young leaves. Several investigators (14–16) have shown that after five or six passages of massive amounts of zoospore inoculum through tomato leaves, potato races can be converted to races with high virulence to tomatoes. Turkensteen (16) demonstrated that this adaptation to tomato occurred naturally in the field when small plots of tomatoes were grown near heavily infected potatoes. In the laboratory, Turkensteen (16) also demonstrated adaptation of potato races of *P. infestans* to *Solanum dulcamara* (bitter nightshade). The initial potato isolates caused no lesions at all on young, intact leaves of *S. dulcamara* but produced small lesions on senescent leaves of *S. dulcamara*. Sporulation on *S. dulcamara* could be induced only by incubating the lesions in contact with sliced potato tubers. After three passages through the senescent leaves, an isolate highly virulent to *S. dulcamara* was obtained. A highly virulent isolate was also obtained beginning with spores from one lesion on a young *S. dulcamara* leaf with a broken midrib.

The implications of these studies for the use of plant pathogenic fungi as classical biocontrol agents are clear. Suppose that the potato plants were the weeds to be controlled and that *P. infestans* was the fungus introduced to control them. Initial screening tests would have indicated that *P. infestans* attacked many species in the Solanaceae but was only weakly virulent on tomatoes. On the basis of this observation, the risk of damage from the fungus might have been misinterpreted. Therefore, when a facultative parasite like *P. infestans* is considered for introduction into a new area of the world, it is important that it be avirulent on useful plants, not just weakly virulent. The introduction of facultative parasites to control weeds closely related to crop plants or valuable native species should not be attempted without extensive studies of the potential for adaptability of the pathogen to new hosts.

There are many other cases in which the ability of a facultative parasite or a facultative saprophyte to attack a new host seems to be under simple genetic control. The most extensive studies of such cases were done by Nelson and Kline (17–20) with *Cochliobolus* spp. They found that isolates of *C. carbonum* differed widely in their host ranges and that virulence to each of 22 different grass species in 21 different genera is controlled by one or two genes. In *C. heterostrophus* they found that virulence to nine different grass species is controlled by one or two genes in each case. Hamid and Aragaki (21) showed that two single genes in *Setosphaeria turcica* control virulence to corn and sorghum. Yaegashi (22) found that virulence of *Pyricularia* to weeping lovegrass (*Eragrostis curvula*) and finger millet (*Eleusine coracana*) is controlled by a single gene in each case. Goldschmidt (23) found that virulence of *Ustilago violacea* to several different related host species is under simple genetic control. In none of these examples were mutations to virulence to a new host actually demonstrated. Nevertheless, the examples do serve to illustrate that there may be serious hazards inherent in the use of these or other facultatively parasitic fungi as classical biocontrol agents.

Apple scab caused by *Venturia inaequalis* provides another example from which to assess the potential hazards of genetic variation in fungi used as classical biocontrol agents. *Venturia inaequalis* appears to be generally adapted to *Malus* spp. Wild species of *Malus* have specific genes for resistance not known to occur in domestic apple cultivars, and these genes appear to be part of a single gene-for-gene system (24). Therefore, it would not be surprising to find mutants of *V. inaequalis* able to attack *Malus* spp. that are resistant to the parent isolate.

Ellingboe (25) suggested that in host-pathogen systems with gene-for-gene specificity, the specific genes for virulence evidently occur in addition to other genes that condition a basic compatibility between host and pathogen. This finding implies that the host range of a pathogen will be determined by many genes with positive effects to adapt the pathogen's growth and reproduction so that they are compatible with the host's metabolism, structure, and life cycle. Superimposed on this basic adaptation are the effects of specific genes for resistance in individual host species and specific genes for virulence in the pathogen to overcome the effects of these genes for resistance.

This seems to be a reasonable theory to guide us in the selection of classical biocontrol agents among plant pathogens. We should consider the breadth of a pathogen's host range as well as the results of pathogenicity tests on specific hosts. If the cultivated species and valuable native wild plant species appear to fall within the general host range of the candidate pathogen, it is possible that the pathogen is already preadapted to these plant species even if the specific isolates tested did not seriously affect the specific host cultivars or wild varieties that were inoculated. If the pathogen is preadapted to a valuable plant species, there is a real danger that a single genetic change might allow it to attack the new host. This is one of the reasons why Watson and Harris (26) had reservations about attempting to use plant pathogens for biocontrol of wild oat in Canada. Even if the pathogen were not virulent to domesticated oat cultivars when biocontrol was initiated, there would be a hazard that new mutations to virulence might appear in the increased population of the pathogen.

HAZARDS OF GENETIC HETEROGENEITY—MICROBIAL HERBICIDES

Most of the potential hazards of classical biocontrol agents can be avoided with microbial herbicides by selecting pathogens that are already endemic in the area where they are to be used. Microbial herbicides may employ pathogens that do not ordinarily kill the target weed or severely limit its reproduction. By applying the pathogen in large doses early in the growing season, the conditions of a severe, local epidemic can be produced before the weedy plant has a chance to mature (2–4). In this way, the severe effects of the biocontrol agent are restricted to the area where it is applied and the immediate vicinity. Moreover, if pathogens with limited potential for long range dispersal are chosen, the area of impact can be limited even more effectively.

The hazards of the microbial herbicide to crops and native wild species may derive primarily from the timing and intensity of inoculation rather than from the danger of genetic changes in the pathogen. During their long coexistence with the endemic pathogens, the native plant species have evolved to cope with the genetic variability of the pathogen. Mutations that may occur in the inoculum of a microbial herbicide are not likely to differ from those that have occurred naturally in the pathogen population.

With a microbial herbicide, the opportunities for adaptation by natural selection of the pathogen to nontarget plant species will be limited because each year the inoculum will originate from the same initial source rather than from a naturally selected population from the previous year. In the case of *Phytophthora infestans* on tomato, Mills (14) found that the fungus overwintered in potato tubers but not in tomato tissue. Each year the tomato race was selected from the pathogen population that developed in potato fields. Because the tomato race developed only after the potatoes were heavily infected, this race did not become established in potato tubers. Therefore, the frequency of the tomato race remained low. Turkensteen (16) suggested a similar relationship of the tomato race to potatoes in the Netherlands. With a microbial herbicide, the pathogen would begin each crop season as a population adapted to the target weed just as *P. infestans* starts as a population adapted to potatoes rather than tomatoes.

For the reasons just described, genetic variability in the pathogen used as a microbial herbicide should pose little or no hazard, especially if the microbial herbicide is used against annual plants in a temperate climate and has limited ability to survive in soil. With microbial herbicides used in tropical climates or against stems or roots of perennial plants, the chances for independent adaptation of the pathogen would be somewhat greater. However, if this hazard were very great, it should have already become apparent as cases of epidemics of cultivated crops spreading into native plants. I know of no cases in which an indigenous plant species has been seriously damaged in this way. Instead, the severe epidemics, such as those of chestnut blight, Dutch elm disease, and white pine blister rust, have all been caused by introduced pathogens. The greatest hazard in the use of microbial herbicides is probably not that of genetic variability, but rather the danger that the pathogen may be deliberately disseminated widely as a microbial herbicide and, consequently, introduced into regions where it did not previously occur.

BENEFITS OF GENETIC HETEROGENEITY—CLASSICAL BIOCONTROL

The primary benefit of genetic heterogeneity in pathogens to be used as classical biocontrol agents has already been illustrated in the case of skeletonweed rust. Because the pathogen population was genetically heterogeneous, it was possible to find an isolate with the desired virulence even though the first collections were completely unsatisfactory. When a pathogen has an extensive geographic range, it is likely that local populations will differ from one another. Each population will be adapted to the climatic conditions and the host population in its local habitat. In the case of skeletonweed rust, if the pathogen had been intended for use in a cooler climate, it would have been logical to begin collection in the more northern parts of the range of the fungus in eastern Europe. Similarly, isolates virulent to a particular form of a weed should be sought where that form is most common.

Hosts of rust fungi or other obligate parasites may occur in a variety of vertical resistance genotypes, as is the case with wild oat in Israel (27) and wild sunflowers in North America (28). If a single strain of the pathogen virulent on all the weed's genotypes could not be found, it could be produced by crossing appropriate pathogen isolates to combine the desired genes for virulence. Procedures for crossing rusts, smuts, powdery mildews, and many other pathogens have been developed in research on the genetics of pathogens of cultivated crops (29).

BENEFITS OF GENETIC HETEROGENEITY—MICROBIAL HERBICIDES

Plant pathogens used as microbial herbicides can be treated as domesticated organisms. Like crop species, they can be bred and adapted for specialized purposes without regard for their ability to compete and survive on their own in a natural ecosystem. For instance, a pathogen that kills its host quickly but cannot compete with saprophytes in the dead tissue might not be successful in nature, but it could be an excellent microbial herbicide.

The processes used to increase the virulence of pathogens and to make them more suitable for mass culture can be similar to those used with microorganisms in the fermentation industry. The first and still most widely used technique for increasing antibiotic production in microorganisms is the induction of mutants. Stepwise increase of antibiotic production by mutation has been so successful that the industry has been slow to adopt the very powerful genetic engineering techniques now available in microbial genetics and molecular biology (30).

The work in plant pathology with mutations to specific (vertical) virulence in rust fungi has already been mentioned. Studies have also been done with similar mutations to virulence in other plant pathogens (29), but there has been little research on increasing levels of polygenically controlled general virulence by induced mutations. Gabriel et al. (31) demonstrated that virulence in *Phyllosticta maydis* can be increased by mutation. They induced six mutants that were more virulent than the original strain at 30°C and four mutants that were more virulent at both 22° and 30°C.

Although several effective chemical mutagens are available, the preferred mutagen for most microorganisms should be ultraviolet (uv) light (32). Ultraviolent light is effective, easy to contain, and cannot be accidentally inhaled or ingested. On the other hand, chemical mutagens are either known or suspected carcinogens. With organisms that are unsuitable for uv irradiation, ethyl methanesulfonate is preferred (32).

It is commonly recognized that populations of plants vary in levels of polygenically controlled general resistance, but it is equally true that populations of plant pathogens vary in polygenically controlled virulence. For example, among 79 isolates of *Cylindrocladium crotalariae*, Hadley et al. (33) found a nearly normal distribution of virulence to two peanut (*Arachis hypogaea*) cultivars in greenhouse tests (Fig. 1). On the resistant cultivar, the level of disease was much reduced, but the most virulent isolate of the pathogen caused nearly as much root rot as the mean for all isolates on the susceptible cultivar.

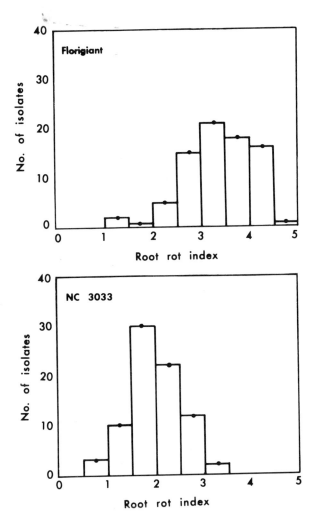

FIGURE 1. Root rot indexes of peanut cultivars Florigiant (susceptible) and NC 3033 (moderately resistant) inoculated with 79 isolates of *Cylindrocladium crotalariae* at a rate of 35 microsclerotia per gram of soil. Root rot indexes range from 0 (healthy, no lesions) to 5 (plant dead, root system completely decayed). Data for each isolate represent the means of five replications with two plants per pot in each replication. Data from Hadley et al. (33).

107

The existence of quantitative genetic variation for virulence in pathogen populations permits the selection of isolates with even greater virulence through recombination of existing genes. Nelson et al. (34) showed that by alternate cycles of crossing and selection, in 3 years it was possible to produce strains of *Setosphaeria turcica* virulent on corn cultivars that were resistant to all the original parent isolates. In these tests the evaluation of virulence was qualitative and based on lesion type, but it is known that the resistance of the corn cultivars was polygenic.

Emara and Sidhu (35) showed that general virulence (aggressiveness) in *Ustilago hordei* is polygenically inherited with a heritability of 65%. Hill (36) obtained estimates of heritability of 21–70%, 23–56%, and 0–13%, respectively, for infection efficiency, sporulation, and lesion size for *Cochliobolus heterostrophus* on corn. In *Aspergillus nidulans* with heritability estimates in this general range, Merrick (37) was able to double the production of penicillin in four or five cycles of crossing and selection. By intercrossing improved lines of the fungus and continuing selection, he was able to raise penicillin production to three times the original level. Similar gains in virulence in *U. hordei* and *C. heterostrophus* should be possible because the heritability of virulence in these fungi is similar to the heritability of penicillin production in *A. nidulans* and because the same principles of genetic variation apply in each case. Furthermore, it would be surprising if similar gains could not be made with most sexually reproducing plant pathogens.

Caten and Jinks (38) have outlined the procedures and methods of analysis employed in studies of quantitative inheritance in fungi. For fungi without sexual stages, it may be possible to produce genetic recombination through the parasexual cycle. Calam et al. (39) used parasexual recombination with *Penicillium chrysogenum* to increase sporulation in strains that yielded high concentrations of penicillin. The use of parasexual recombination is more difficult than sexual recombination (when it is available), and the results are less predictable. Nevertheless, Calam et al. (39) were able to make significant progress in improving their strains of *P. chrysogenum*.

In the use of microbial herbicides as well as chemical herbicides, there is a potential problem that the resistance of the weeds may increase through natural selection (see Chapter 6). The only recourse with chemical herbicides is to increase the dosage, which may be unacceptable, or to seek a new herbicide. With microbial herbicides, there is a third alternative. The pathogen can be selected periodically on the weed population to increase its virulence to match the increasing resistance in the plant host. Research by Bausback (40) on ergot of rye provides experimental support for this prospect. Bausback selected simultaneously for increased resistance to ergot within a population of rye and for increased virulence within a population of the ergot fungus *Claviceps purpurea*. After four cycles of selection, both resistance and virulence had significantly increased, but the level of disease caused by the selected pathogen genotypes on the selected host genotypes was almost exactly the same as the level of disease caused by the unselected pathogen on the unselected host.

CONCLUSIONS

In comparing the potential hazard of genetic heterogeneity in *Puccinia chondrillina* and the hypothetical hazards with *Phytophthora infestans*, *Cochliobolus* spp., and *Venturia inaequalis*, it is apparent that simple host range tests with one or two isolates of the pathogen and a single genotype of each host do not adequately define the pathogenic potential of the fungus. Some hosts that appear resistant in such tests may become seriously diseased if exposed to the pathogen through several disease cycles. In other cases, a host that appears slightly susceptible in the first screening test may never be significantly damaged by the pathogen during a long association with it. Therefore, with pathogens of the classical biocontrol type, it is important to obtain evidence on potential hazards to crop plants in the country of origin of the pathogen. Most major crops are widely grown around the world and can be inspected for any signs of damage from the candidate pathogen. The hazards to native wild species from an imported pathogen are not so easily evaluated. In this case, good judgment and a thorough knowledge of plant pathology and the nature of related pathogens are required. In general, importation of plant pathogens to control weed species closely related to desirable native species is risky. If importation is undertaken, it should be preceded not only by an extensive host range test, but also by adaptation tests similar to those described for *P. infestans* on tomato.

The potential benefits of genetic manipulations of plant pathogens used as classical biocontrol agents may be limited because plant pathogens of this type must retain their natural competitive and survival abilities. With excessive genetic manipulation, they are likely to become like cultivated crops with qualities that render them highly productive when they are under human care but not very productive or successful when left to survive on their own. As Mayr (41) stated, there are no all-purpose genotypes. "All fitness is a compromise, and gaining an advantage in one way usually means abandoning some other properties. . . ."

With microbial herbicides, the potential for genetic improvement is much better because the pathogens can be adapted for specialized purposes. The great progress made in selecting for specific attributes in industrial microorganisms exemplifies what is yet to be accomplished with microbial herbicides.

REFERENCES

1. J. R. Williams. 1954. The biological control of weeds. Pp. 95–98 in: *Report of the Sixth Commonwealth Entomological Conference.* Commonwealth Institute of Entomology, London.
2. G. E. Templeton and R. J. Smith, Jr. 1977. Managing weeds with pathogens. Pp. 167–176 in: J. G. Horsfall and E. B. Cowling, eds., *Plant Disease: An Advanced Treatise.* Vol. 1. *How Disease is Managed.* Academic Press, New York.
3. G. E. Templeton, D. O. TeBeest, and R. J. Smith, Jr. 1979. Biological weed control with mycoherbicides. *Annu. Rev. Phytopathol.* **17**, 301–310.
4. H. D. Ohr. 1974. Plant disease impacts on weeds in natural ecosystems. *Proc. Am. Phytopathol. Soc.* **1**, 181–184.
5. P. Harris. 1973. The selection of effective agents for the biological control of weeds. Pp. 75–85 in: A. J. Wapshere, ed., *Proceedings of the Third International Symposium on Biological Control of Weeds.* Commonwealth Agricultural Bureaux, Slough, England.
6. S. Hasan. 1972. Specificity and host specialization of *Puccinia chondrillina. Ann. Appl. Biol.* **72**, 257–263.
7. J. M. Cullen, P. F. Kable, and M. Catt. 1973. Epidemic spread of a rust imported for biological control. *Nature* **244**, 462–464.
8. J. M. Cullen. 1973. Seasonal and regional variation in the success of organisms imported to combat skeletonweed *Chondrilla juncea* L. in Australia. Pp. 111–117 in: A. J. Wapshere, ed., *Proceedings of the Third International Symposium on Biological Control of Weeds.* Commonwealth Agricultural Bureaux, Slough, England.
9. I. A. Watson. 1957. Mutation for increased pathogenicity in *Puccinia graminis* var. *tritici. Phytopathology* **47**, 507–509.
10. D. E. Zimmer, J. F. Schafer, and F. L. Patterson. 1963. Mutations for virulence in *Puccinia coronata. Phytopathology* **53**, 171–176.
11. H. H. Flor. 1958. Mutation to wider virulence in *Melampsora lini. Phytopathology* **48**, 297–301.
12. G. J. Green. 1971. Hybridization between *Puccinia graminis tritici* and *Puccinia graminis secalis* and its evolutionary implications. *Can. J. Bot.* **49**, 2089–2095.
13. T. Johnson and M. Newton. 1946. Specialization, hybridization, and mutation in the cereal rusts. *Bot. Rev.* **12**, 337–392.
14. W. R. Mills. 1940. *Phytophthora infestans* on tomato. *Phytopathology* **30**, 830–839.
15. K. M. Graham, L. A. Dionne, and W. A. Hodgson. 1961. Mutability of *Phytophthora infestans* on blight-resistant selections of potato and tomato. *Phytopathology* **51**, 264–265.
16. L. J. Turkensteen. 1973. *Partial Resistance of Tomatoes Against Phytophthora infestans, the Late Blight Fungus.* Agricultural Research Report 810, Institute for Phytopathological Research, Wageningen, The Netherlands, 88 pp.

17. R. R. Nelson and D. M. Kline. 1963. Gene systems for pathogenicity and pathogenic potentials. I. Interspecific hybrids of *Cochliobolus carbonum* × *Cochliobolus victoriae*. *Phytopathology* 53, 101–105.

18. R. R. Nelson and D. M. Kline. 1969. The identification of genes for pathogenicity in *Cochliobolus carbonum*. *Phytopathology* 59, 164–167.

19. R. R. Nelson. 1970. Genes for pathogenicity in *Cochliobolus carbonum*. *Phytopathology* 60, 1335–1337.

20. R. R. Nelson and D. M. Kline. 1969. Genes for pathogenicity in *Cochliobolus heterostrophus*. *Can. J. Bot.* 47, 1311–1314.

21. A. H. Hamid and M. Aragaki. 1975. Inheritance of pathogenicity in *Setosphaeria turcica*. *Phytopathology* 65, 280–283.

22. H. Yaegashi. 1978. Inheritance of pathogenicity in crosses of *Pyricularia* isolates from weeping lovegrass and finger millet. *Ann. Phytopathol. Soc. Jap.* 44, 626–632.

23. V. Goldschmidt. 1928. Vererbungsversuche mit den biologischen Arten des Antherenbrandes (*Ustilago violacea* Pers.). Ein Beitrag zur Frage der parasitären Spezialisierung. *Z. Bot.* 21, 1–90.

24. D. M. Boone. 1971. Genetics of *Venturia inaequalis*. *Annu. Rev. Phytopathol.* 9, 297–318.

25. A. H. Ellingboe. 1976. Genetics of host-parasite interactions. Pp. 761–778 in: R. Heitefuss and P. H. Williams, eds., *Physiological Plant Pathology. Encyclopedia of Plant Physiology, New Series.* Vol. 4. Springer-Verlag, West Berlin, West Germany.

26. A. K. Watson and P. Harris. 1975. Weed control with plant pathogens and nematodes. *Can. Agric.* 20 (4), 26–27.

27. I. Wahl. 1970. Prevalence and geographic distribution of resistance to crown rust in *Avena sterilis*. *Phytopathology* 60, 746–749.

28. D. E. Zimmer and D. Rehder. 1976. Rust resistance of wild *Helianthus* species of the north central United States. *Phytopathology* 66, 208–211.

29. P. R. Day. 1974. *Genetics of Host–Parasite Interaction.* Freeman, San Francisco, 238 pp.

30. G. Pontecorvo. 1976. Presidential address. Pp. 1–4 in: K. D. Macdonald, ed., *Second International Symposium on the Genetics of Industrial Microorganisms.* Academic Press, London.

31. D. W. Gabriel, A. H. Ellingboe, and E. C. Rossman. 1977. Mutations affecting virulence in *Phyllosticta maydis*. *Can. J. Bot.* 57, 2639–2643.

32. B. A. Bridges. 1976. Mutation induction. Pp. 7–14 in: K. D. Macdonald, ed., *Second International Symposium on the Genetics of Industrial Microorganisms.* Academic Press, London.

33. B. A. Hadley, M. K. Beute, and K. J. Leonard. 1979. Variability of *Cylindrocladium crotalariae* response to resistant host plant selection pressure in peanut. *Phytopathology* 69, 1112–1114.

34. R. R. Nelson, A. L. Robert, and G. F. Sprague. 1965. Evaluating genetic potential in *Helminthosporium turcicum*. *Phytopathology* 55, 418–420.

35. Y. A. Emara and G. Sidhu. 1974. Polygenic inheritance of aggressiveness in *Ustilago hordei*. *Heredity* 32, 219–224.

36. J. P. Hill. 1978. The heritability and genetic control of certain attributes of parasitic fitness of *Helminthosporium maydis* Race T. Ph.D. thesis. Pennsylvania State University, University Park, PA, 59 pp.

37. M. J. Merrick. 1976. Hybridization and selection of penicillin production in *Aspergillus nidulans*—a biometrical approach to strain improvement. Pp. 229–242 in: K. D. Macdonald, ed., *Second International Symposium on the Genetics of Industrial Microorganisms*. Academic Press, London.

38. C. E. Caten and J. L. Jinks. 1976. Quantitative genetics. Pp. 93–111 in: K. D. Macdonald, ed., *Second International Symposium on the Genetics of Industrial Microorganisms*. Academic Press, London.

39. C. T. Calam, L. B. Daglish, and E. P. McCann. 1976. Penicillin: Tactics in strain improvement. Pp. 273–287 in: K. D. Macdonald, ed., *Second International Symposium on the Genetics of Industrial Microorganisms*. Academic Press, London.

40. G. A. Bausback. 1976. Kunstliche Selektion in dem Wirt-Parsit-System Roggen-Mutterkorn. Ph.D. thesis. University of Hohenheim, Stuttgart, West Germany, 85 pp.

41. E. Mayr. 1965. Summary. Pp. 553–562 in: H. G. Baker and G. L. Stebbins, eds., *The Genetics of Colonizing Species*. Academic Press, New York.

8

CREATING EPIPHYTOTICS

ROBERT D. SHRUM

U.S. Department of Agriculture, Science and Education Administration, Agricultural Research, Plant Disease Research Laboratory, P. O. Box 1209, Frederick, Maryland

Biological pest control methods are based on natural systems, and natural ecosystems represent the epitome of efficiency and conservation of energy. Through the process of evolution, each available ecological niche is occupied by the most efficient and effective process available, and this process remains until a better one displaces it. The most efficient one survives. Humans can profit from that long experience in trial and error by choosing from nature's list of surviving control systems for their own pest control needs. From this perspective, every fungus, bacterium, virus, and other micro- and macroorganism pathogenic to weeds is a potentially useful weed control agent. From a biocontrol perspective, the determination as to whether the biocontrol agent's demonstrated pathogenic capability can be made safe for crops and sufficiently destructive to weeds is in several parts, two of which are of concern in this chapter: (1) the recognition of a superior epidemiologic potential of a pathogen and (2) the determination as to whether that potential can be used for timely pest control. The principles of epidemiology set the base for understanding the former, and in conjunction with information on how the duration and intensity of disease mediate host damage, suggest solutions for the latter. This chapter outlines some of these basic considerations.

This chapter addresses ways to create epiphytotics (plant disease epidemics). Underlying the chapter title is a series of important questions. Can epidemics that are effective in reducing weed populations be created by humans, and if so, what must be done to promote these epidemics in weed populations? What is an effective epidemic? What level of disease might give effective weed control? What are the criteria for judging effectiveness of weed pathogens, and might there be more than one set of standards for a given biocontrol situation?

USE OF PLANT PATHOGENS FOR BIOLOGICAL WEED CONTROL

The potential use of plant pathogens as biological weed control agents is relatively unexplored. Until recently, epidemics on weeds have been of peripheral concern to epidemiologists. Although most of these epidemics pose no direct threat to crops, they have potential value in weed control. Weed control has been the established domain of weed scientists, not pathologists. Weed scientists are seldom fully trained in pathology, and pathologists have been professionally inhibited from working in weed control. Thus this potentially fruitful area, bridging the two disciplines, has been relatively unattended for a long time. Although the situation is changing, a recently published textbook on modern weed control (1) points to the continuing problem. This text lacks any reference to the use of fungal and bacterial pathogens as weed control agents. Fungal, bacterial, and viral pathogens are still being overlooked as components of modern weed control systems. Plant pathologists must accept some responsibility for this neglect. Of nearly 600 papers presented at the 1980 Annual Meeting of the American Phytopathological Society, only two presentations dealt with the use of plant pathogens as weed control agents. This inattention seems to have its base more in plant pathological tradition than in reasoned rejection of pathogens for biocontrol of weeds.

Clearly, the focus of plant pathology has been on disease prevention on crop plants, whereas weed control with plant pathogens would require an effort to promote plant disease, not retard it. This traditional view is changing, and many crop scientists now recognize that the principles developed for crop pathology will also be useful in biocontrol applications. The same explosive growth of microorganisms that can render pathogens dangerous to crops can make them very useful in the control of weeds. We must integrate nature's time-tested systems with our own to maximize pest control capabilities.

IMPROVING NATURE'S BEST

The pace of the reproductive cycle is an important aspect of population dynamics. Organisms that have rapid-paced reproductive cycles ultimately might be the most effective biocontrol agents because a potential for large populations is ensured. In general, the capability for more reproductive cycles during a given period imparts a potential for more rapid buildup of the pathogen. Whether the rapid buildup translates into better weed control depends on many other factors as well, but rapid-paced reproduction is a prime factor in the epidemiologic potential of a pathogen.

Rapid-paced reproduction is common to many agricultural pests. Organisms such as weeds would not be menacing to crop growth if they lacked relatively faster cycles than those of the crops they menace. By selecting biocontrol agents with the quickest life cycles such as fungal or viral pathogens, the selective advantage of quick life cycles of the weeds would be lessened. The time taken for weeds to complete a reproductive cycle normally ranges from weeks to months. Viruses are capable of reproducing within minutes to hours. Thus, theoretically they can complete thousands of cycles during one life cycle of the weed, with each new generation multiplying manyfold over the previous. Fungi can reproduce within days to weeks and bacteria in hours to days, giving the potential for up to tens or hundreds of cycles, respectively, during a single life cycle of the weed. Therefore, the potential reproductive capabilities of plant pathogens can be staggering.

The advantage given by the rapid-paced reproduction of plant pathogens can be further magnified through management strategies that promote population increases by removing impediments to the normal development of the pathogens. Just as crop plants are selectively propagated by collecting seeds prior to an inclement season and redisseminating them during periods favorable for growth, various pathogens can be promoted on a regular basis, beyond where they would normally develop, by collecting their propagules and disseminating them to susceptible weed populations when conditions most favor infection and disease development. Because rapidly reproducing microorganisms can reach explosive population levels, even a minor buildup of propagules early in the season can shift natural balances so that the pathogens will overwhelm their hosts. Thus the more explosive the population of the organism involved, the greater will be the advantage predestined by the propagation effort. More discussion is presented on this later in this chapter.

DESCRIBING THE OBVIOUS

A textbook on ecology by Daubenmire (2) claimed, "ecology is nothing more than an elaboration of the obvious." I support this contention, but stress that the implication is not that the various ecological systems are

necessarily simple. Agroecosystem complexity is somewhat facetiously illustrated by Fig. 1, where the arrows and boxes represent cause and effect
relationships. The diagram is complex and messy, and although it may be
annoying to follow, none of the diagrammed relationships is likely to be
novel or surprising to any serious agriculturist. Part of the intent of Fig. 1
is to show that the process of evaluating a functioning system must include
operations other than simply determining what the interactions are
and/or the quantitative nature of those interactions. For overall understanding and for forecasting, the form of information organization can be
as important as the information itself. Organization based on a logical
structure, and on generic categories of information can greatly simplify
the system without loss of information or detail. In the discussion that fol-

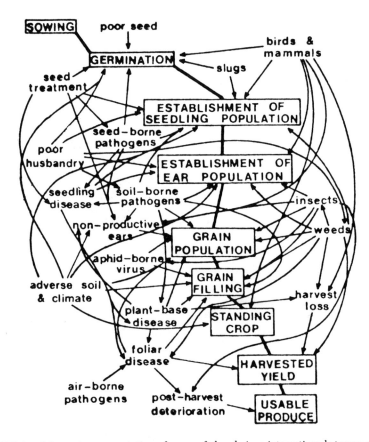

FIGURE 1. Schematic representation of some of the obvious interactions between certain
factors that affect cereal crop production. Courtesy of Richardson (3).

lows, a sequence of diagrams is used to develop such a set of categories to gain a perspective on the epidemic dynamics to be discussed later. This perspective is necessary to utilize in a specific way the general principles developed later.

OVERVIEW OF INTERACTIONS

There are many alternative ways to represent interactions of the agro-ecosystem. Figure 1 is only one such representation—not particularly functional for our purposes. Figure 2a separates those components of greatest interest for biological weed control. Here the crop plant, the weed pest, and the environment become the focus of interest on opposing tips of a triangle. Figure 2b adds the biocontrol component as well and illustrates that all three biological entities must function within the same environment. Figure 3a takes this sequence one step further to illustrate that although the weed (host), biocontrol agent (pathogen), and crop function within the same overall environment, each is probably functionally related to different components of that environment (4). Some stimuli affect one organism more than others, and still other organisms are not affected at all.

For the purpose of this discussion, environment can be viewed in the manner of Platt and Griffiths (4). Only that portion of the overall environment that is functionally related to the various participants will be of interest. The weed shown in Fig. 3a may be exposed to various physical factors such as wind and leaf wetness, which may have no discernible impact on its growth. By definition, lack of a functional impact removes these factors from the weed's functional environment. On the other hand, if in any way the factors directly affect the pathogen, they remain a part of its environment. Thus phytophagous insects may be a part of the functional environment of the weed although not directly that of the pathogen. Figure 3b further simplifies this ecosystem representation. For simplicity, only the direct interactions between the weed host, pathogen, and their respective environments are dealt with in this chapter. Any number of indirect and collateral interactions can also be represented without greatly increasing complexity under this organizational arrangement. For example, all agents that remove leaf tissue (insects, physical forces, herbivorous animals, or senescence) are included under one category—reduction of host tissue. With this arrangement, specific causes are summarized into a single inclusive term. Collateral effects are lumped together with other sources of that same effect into terms representing other direct impacts. This summarized form has biological meaning while

Crop Production Triangle

(Disease Triangle)

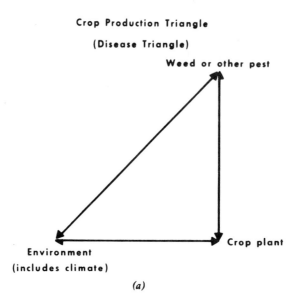

(a)

Biocontrol
Double Triangle

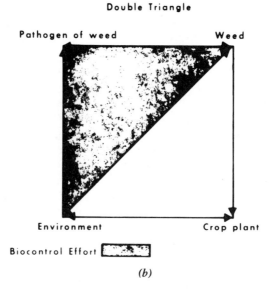

(b)

FIGURE 2. Schematic representation of the interplay of environment with pest, crop, and biocontrol agent: (*a*) in a crop system (typically crop production is less than optimal as a result of various pests and less than favorable environments); (*b*) in a biocontrol system where the shaded area represents the addition of a biocontrol agent to the crop system.

118

SYSTEM STRUCTURE

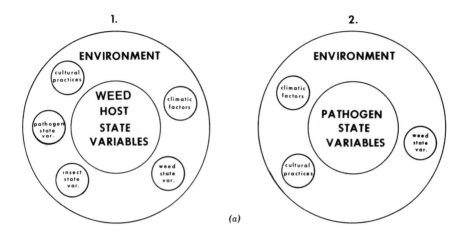

(a)

SYSTEM FUNCTION

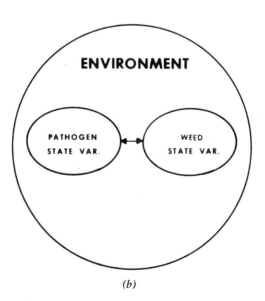

(b)

FIGURE 3. Relationship of system structure to system function: (a) state of the weed host and state of the pathogen are affected by their respective environments; (b) the pathogen is the major component of the host's state, and vice versa.

119

avoiding some of the confusion illustrated in Fig. 1. Organizationally, the interacting host and pathogen become codependents, responding to each other and to the rest of their environment by the physical stimuli that they exchange—regardless of whether the actual source of a stimulus was multiple or single, or physical or biotic in nature. As response is determined by the size of the various stimuli, the influence of any management activity can be evaluated by judging its contribution to the various aspects of each organism's functional environment. Determination of how specific ecosystem forces contribute to any particular stimulus is pretty much left to the reader, however.

EPIDEMIC DYNAMICS

Epidemiology is a quantitative science, but no attempt is made here to be unequivocally precise or mathematical. The prime concern is to make current epidemiologic knowledge serve the biocontrol setting and to illustrate basic principles of pathogen population dynamics. Both may be done with minimum mathematical manipulations.

CREATING AN IMBALANCE

Native pathogens seldom cause raging epidemics in natural settings. To understand how epidemics can be created on weeds, it is first necessary to know why efficient organisms (see introductory paragraph) do not flourish naturally and then proceed to eliminate or bypass the major impediments to their development (see also Chapter 5). To survive within natural ecosystems, weeds and their native pathogens must develop mechanisms that keep them in balance with each other. Selection pressures over evolutionary time have created a near steady-state situation—a natural balance. Without such balances, we would either not know the pathogen or not have to deal with the weed; one or the other would have been eliminated. This natural balance is usually maintained by the following three general categories of constraints:

1. Innate resistance of the weed host (genetic resistance).
2. Environmental deficiencies (environmental resistance).
3. Lack of inoculum (isolation in time or space from sources of inoculum):
    a. With exotic pathogens spatial isolation is severe (the pathogen gets no opportunity to attack its host).

   b.   With native pathogens (endemics):

   (i)   Depending on the degree of spatial discontinuity of the host, spatial isolation is sometimes mild and sometimes severe.

   (ii) Inoculum often survives severe overwintering and/or oversummering conditions at critically low levels. This can lead to spatial discontinuity of inoculum and thus greater isolation of the host from the pathogen.

If epidemics do *not* occur in the field, it is impossible to determine which of these three types of pressure has contributed most to their absence. Genetic and environmental resistance are commonly assumed to be responsible, and hence the importance of the inoculum-reducing processes of item 3 in the preceding list are often overlooked. The largely untested assumption that weeds are resistant to native pathogens underlies that bias.

Undoubtedly, some weeds have developed some resistance to some native pathogens, but the assumption that exotic pathogens will work better, in a general sense, is premature. Many native pathogens have the genetic potential to be as successful as exotics; they need only to be enhanced by proper methods. This should not be particularly surprising to crop pathologists. It is well known that most crop diseases are not epidemic in most years; yet with a single, well-timed application of inoculum, such epidemics can be successfully created about 80–90% of the time. In a 6-year study at eight Midwestern locations, 100 of 111 test plots of wheat inoculated with wheat stem rust had significantly greater amounts of disease and greater yield reductions than did plots without inoculation (5). The remaining plots had significantly greater amounts of disease than did uninoculated areas, even though yield reductions were not significantly different. Multiple inoculations would have ensured even greater differences. Testimony can be obtained from almost any researcher for any of a number of native pathogens that timely inoculations can routinely cause epidemics whereas otherwise they would seldom occur. Thus neither were the hosts too resistant, nor the natural environment too restrictive, but the epidemics were prevented or retarded because inoculum was too deficient. Weeds and weed pathogens exist under similar circumstances, and many of these pathogens have the additional problem created by the lack of spatial continuity of their hosts. Even with this additional impediment, during certain years some patches of weeds do become severely diseased by native pathogens. Often such patchy occurrence is proposed as proof of genetic diversity of the weed (i.e., resistance) when in fact it could just as well be the consequence of low inoculum and/or spatial isolation.

One may wonder justifiably whether genetic resistance of the host slows the increase of disease or the lack of disease, due to other reasons, slows the increase of host resistance. The latter half of the question seems as appropriate as the first half. Low postwinter inoculum levels and spatial isolation lessen selection pressures for host resistance. The section on mathematics of management actions shows how the unusually low postwinter inoculum levels minimize selection pressures. Additionally, as discussed in Chapter 6, the genetic systems of many weeds are well buffered from change through seed or rhizome survival and by spatial isolation. I suspect that the significance of host resistance and the role of genetic diversity in preventing epidemics by native pathogens has been overestimated. This hypothesis needs testing, however.

Native pathogens are locally available and locally adapted, and to some extent local agricultural crops have been successfully screened against their pathogenic capabilities. These advantages alone perhaps provide sufficient motive for initiating more extensive epidemiologic testing. Additionally, testing of native pathogens as weed biocontrol agents can provide an opportunity for evaluating the principles of creating epidemics.

INFECTION WINDOWS

An infection window is the time during which a weed host is susceptible and environmental conditions are favorable for infection. Although individual weeds may grow for many months or even many years, there is normally a limited number of short periods during which an infection window exists. Maximization of the use of these infection windows maximizes the pathogenic potential. Effectively utilizing infection windows is a keystone to creating epidemics.

The first of two factors that determine infection windows is host susceptibility. Some weeds are susceptible only as young plants, some are susceptible as older plants, and some are relatively susceptible throughout their lives. Also, some parts of the host are relatively more susceptible than others. The length of time that the host is susceptible, the timing of inoculation (in the life cycle of the plant), and the degree of susceptibility of the host have major impacts on disease development, and these must be studied to obtain maximum use of infection windows.

Favorable climate is another aspect of the infection window. The physical environment fluctuates hourly, daily, and seasonally. The environmental conditions of spring and fall are normally more suitable for host infection and survival of a pathogen than are summer or winter conditions. Free moisture is commonly required for spore germination, so rain and dew often favor infection, especially when temperatures are moderate. As a result, inoculum produced early in the growing season (spring) is normally more effective than inoculum produced later (summer). Also, daily or seasonal extremes in temperature (highs or lows) can prevent infection, inactivate spores and other infective propagules, and kill or prevent sporulation of established infections. Each pathogenic organism has a specific set of physical requirements. Although most fungal processes function best within temperature thresholds of 0–30°C, with optima around 20°C, pathogens can be found for almost any climate in which weeds are capable of surviving. The pace of pathogen reproduction varies widely, and the rate of reproduction changes almost continuously as various environmental stimuli fluctuate on daily and seasonal cycles.

DISTRIBUTION OF INOCULUM

Epidemics develop by increasing both the intensity of disease at a given site and the number of sites or area of infection. Figure 4 shows consecutive epidemics of various intensities, and Fig. 5 illustrates the spatial spread of infection. Although the basic process for creating epidemics is straightforward, opportunities for creating epidemics are often lost be-

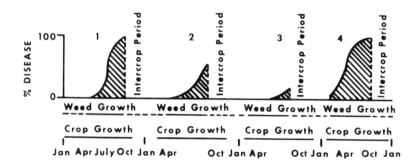

YEAR 1 = LOW INITIAL INOCULUM (i), HIGH RATE (r), INTERMEDIATE TIME (t)

YEAR 2 = LOW i, INTERMEDIATE r, LONG t

YEAR 3 = LOW i, INTERMEDIATE r, SHORT t

YEAR 4 = HIGH i AND r, LONG t

FIGURE 4. Consecutive epidemics of the same organism may vary in impact (shaded areas), according to variation in x_0 (denoted i in this diagram), r, and t (see p. 127).

SPATIAL DYNAMICS

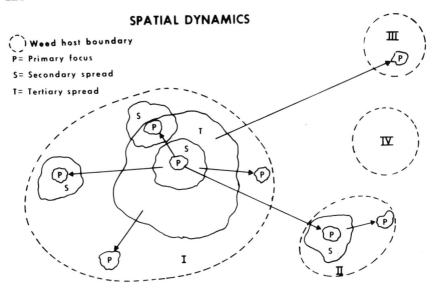

FIGURE 5. Disease increases spatially with overlapping foci of infections. There is a decreasing gradient of disease with increasing distance from the primary focus (P). Disease onset is progressively delayed in outlying areas, and this delay is magnified by host discontinuity or patchiness.

cause inoculum is not distributed sufficiently early or uniformly. This can be illustrated by observing the progressive delays in disease onset in weed patches I, II, III, and IV in Fig. 5. Patch IV still has no infection.

To utilize the genetic capabilities of a pathogen fully, viable inoculum must be applied to the weed population when the weeds are most susceptible and the environment is optimal (minimally, when favorable conditions for infection exist). The term "population" is a keyword in the description. When some weeds become infected and others escape infection, epidemic onset is delayed. Nonuniformity of inoculum distribution in effect isolates the pathogen from part of the weed population, thus delaying inoculum buildup, which has a double impact on epidemic dynamics: (1) delays prevent the opportunity for full utilization of the more numerous infection windows of the early spring and thus more cycles and (2) because the pathogen has missed the spring weather, its fecundity is likely to be lower. Thus the rate of multiplication per infection cycle will be lower, and disease increase will be slower. The relative impacts of such delays are discussed in conjunction with Fig. 6.

When they are sufficiently isolated, weed patches may completely avoid disease throughout the season with obvious consequences. This is an example of the case where host genetic resistance, if it exists, has no opportunity to slow the increase of disease, but spatial isolation will slow the increase of host genetic resistance (see section on creating an imbalance).

The very fact that deficient inoculum is a common early-season condition in natural ecosystems affords humans an ideal opportunity to create the imbalance that can lead to destructive epidemics. Genetic resistance and unfavorable environments are no doubt important in the normal progress of epidemics on weeds, but postwinter inoculum levels are most certainly a prime force maintaining natural balances. Offsetting that balance by making springtime inoculations is perhaps the easiest of the biocontrol management functions to implement.

Nature's interseasonal reduction of inoculum will be more severe for foliar pathogens than for soil-borne pathogens, more typical for obligate than for facultative parasites, more common for annuals and deciduous perennial hosts than for situations where host organs (and thus disease) are not shed during an intercrop (between seasons of weed growth) period, more typical for those pathogens having few resting structures and poor propagule dissemination capabilities than for those with long-term resting structures and/or good dissemination capabilities, and more common for organisms with higher rates of disease increase (r values).

Although some pathogens will be influenced to a greater degree than others, the overwintering and oversummering reductions are almost universal phenomena. To the degree that inoculum reductions are not effected by natural sanitation, selection pressures will increase and the host will either perish or develop resistance and other forms of escape sufficient to maintain a natural balance. A significant aspect of knowing when springtime inoculations can be used to create epidemics is to ascertain the extent to which a given disease will normally retreat during unfavorable weather, and from that, surmise (using Fig. 6) how the disease's normal balance will be affected if this natural impediment is overcome.

To emphasize the importance of intercrop inoculum reductions, it is useful to examine some diseases where inoculum reductions have been relatively ineffectual (which could represent a biocontrol situation in which adequate inoculum would be applied early each growing season). Dutch elm disease, oak wilt, chestnut blight, and various other vascular diseases of perennial hosts avoid severe winter/summer setbacks because their causal organisms survive inclement periods within the buffered environment of living host tissues. Because these pathogens have the innate capability for being destructive (capable of being easily disseminated, aggressive, etc.), and because they are *not* practically eliminated once or twice a year, they cause some of the most destructive epidemics ever known.

All these diseases happen to be caused by exotic pathogens. It would be easy to assume that such destructiveness was associated with the fact that the causal organisms are exotic and thus that their destructiveness was primarily due to a lack of host resistance. That is only part of the story. Obviously, exotic pathogens have been effective because host resistance to those pathogens is limited; however, just as importantly, the pathogen has significant protection from intercrop sanitation. The same host destructiveness would not occur, even for exotic pathogens, if winter sanitation was effective. Even today Dutch elm disease and oak wilt can be effectively controlled by removal of diseased trees (primarily during the winter)—a human-imposed sanitation where natural sanitation was ineffective. Human-imposed sanitation has not worked for chestnut blight because that fungus also survives as a saprophyte at various locations and in species other than diseased chestnuts. Overwintering inoculum cannot be effectively reduced by destroying only the diseased chestnut trees.

The same winter sanitation phenomenon can be demonstrated with many native pathogens and nonperennial hosts as well. Late blight of potatoes and bunt of wheat are cited to make the point. Potato is not grown as a perennial crop, but potato tubers are stored from season to season in protective bins, and by this storage procedure the late blight pathogen *Phytophthora infestans* is protected through the intercrop period. As a result, the late blight pathogen is redisseminated to the field each spring when infected tubers are discarded on cull piles or inadvertently planted with the crop. After nearly a century and a half of destructive epidemics and strong breeding programs to develop blight resistance, the keystone to effective late blight control remains the destruction of infected tubers before they are disseminated in the spring (a human-imposed intercrop sanitation where natural sanitation had been inadvertently bypassed). The same is true for other pathogens that are spared the normal consequences of a highly lethal intercrop environment. Spores of the bunt fungi of wheat (*Tilletia* spp.) are carried from season to season on stored grain. By using effective preplant seed treatments that kill spores carried on the seeds, we have been spared the ravages of bunt epidemics that were once frequent. There are numerous additional examples that could be cited to emphasize this point.

From these examples it can be deduced that for widely distributed perennial weeds there may be a stronger justification to search for exotic biocontrol agents than for less well distributed, nonperennial weeds. Since the former would be widely distributed, they are more readily exposed to native pathogens. Since they have perennial tissues, the pathogen is more likely to be protected from periodic total destruction once the disease has begun. If such hosts had not developed significant resistance to native pathogens long ago, they would no longer exist. Well-distributed perennial hosts are unable to benefit from the inoculum-reducing forces. Therefore, with native pathogens, the natural balances must be maintained by forces such as genetic or environmental resistance. From a weed control perspective, genetic and environmental resistances would be constraints much more difficult to overcome by any type of management activity than constraints due to deficient inoculum. Thus exotic pathogens may be the most suitable biocontrol agents of perennial weeds.

The benefits of disease escape (and thus the lack of selection pressure for disease resistance) will be more fully realized as we proceed in the direction of poorly distributed hosts and/or evanescent tissues. From the perspective of the pathogen, disease escape is less likely as one proceeds in the direction of better nonhost survival and easier long-distance spread of propagules.

MATHEMATICS OF MANAGEMENT ACTIONS

GENERAL MODEL

How can early inoculum and the use of infection windows influence epidemic dynamics? What are the specific effects to be expected with a given disease management action?

All factors that interplay in the dynamics of any disease can be considered in terms of their impacts on three factors: x_0, r, and t, as illustrated in the equation

$$x = x_0 e^{rt}$$

where x is the current disease intensity, x_0 is the initial level of disease or inoculum, e is the base for natural log (a constant), r is the rate of disease increase (reproduction), and t is the duration of the interaction time. *[See van der Plank (6) for more details.] The impact of any management activity can be traced to its effect on one of these three factors. I shall present quantitatively how x_0, r, and t interplay and illustrate qualitatively how various efforts in disease management will affect them and thus affect epidemic dynamics.

---

*Apart from the setting of a single season perspective, a simple interest disease (6) is nothing other than a compound interest disease (6) with a strong (intercrop) sanitation between each infection cycle. For the purpose of this discussion, a distinction between them is not necessary. The stated relationships made here have the same general applications for both types of disease.

Unless otherwise stated, an interaction will be assumed to occur if and only if the host is susceptible, the environment is favorable, and the pathogen is present [i.e., a functional interaction (4)].

By the preceding equation, disease intensity can be determined at any time by evaluating x_0 (the initial level of disease), r (the rate of reproduction), and t (the time of host-pathogen interaction). Thus when an epidemic is acted on by its environment (including management activities), the impact of that environment can be evaluated by determining how x_0, r, and t might be affected. The quantitative relationship of x_0, r, and t to epidemic progress is discussed in a general way. Then specific management actions are discussed in terms of their probable impacts on one or more of those specific population parameters.

The rate of reproduction r is changed by the interplay of various genetic and/or environmental factors for both host and pathogen (represented in Fig. 3b) and by relative spatial isolation of the suscept from sources of inocula. The value of r can be increased in one of five general ways by (1) increasing the number of infections that result from a given amount of inoculum (less resistant weed host, more aggressive pathogen, or more favorable environment), (2) increasing the size of lesions (again due to either the host or pathogen genetic systems or the environment), (3) reducing the length of the latent period (period from when an infection occurs to when new propagules are available for reinfection), (4) increasing the level of propagule production per unit area of infection, and (5) increasing dissemination effectiveness (better spore flight, less spatial isolation, etc). The value of r is seldom greater than $1.0/(\text{unit} \cdot \text{day})$—a value of $0.69/(\text{unit} \cdot \text{day})$ is roughly a doubling of disease per day. [See van der Plank (6) and Populer (7) for further comments.]

The time factor t and initial inoculum levels x_0 are both intimately related to the earliness and the amount of inoculum available for early infection cycles of each new season. These are also strongly influenced by spatial discontinuity of the host, and both are directly related to the degree of intercrop sanitation.

GENERAL EFFECTS

The general interplay of x_0, r, and t is illustrated with a series of graphs in Fig. 6, for which time is represented along the abscissa and disease intensity (severity) on the ordinate. In Fig. 6 the relative effect of increasing initial inoculum (x_0) is shown as one progresses from the top to the bottom row of graphs. The time factor t increases from left to right, and different reproductive rates r are illustrated by the three individual curves within each of the nine sets of graphs. These 27 curves show the full range of

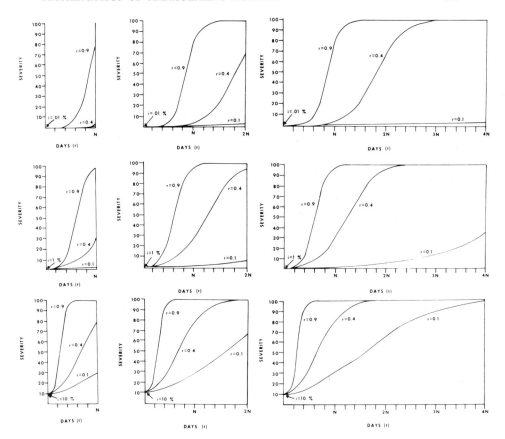

FIGURE 6. As time t (in days) increases from left to right and initial inoculum x_0 (denoted i in this diagram) levels increase from top to bottom, the three different disease rates r in each graph illustrate a range of disease scenarios. See also Table 1 in this chapter.

epidemic response to combinations and permutations of the three basic factors that effect changes in disease intensity. For example, the graph at row 3 (bottom), column 3 (right) shows disease progress curves for three different r values (0.9, 0.4, and 0.1) corresponding to the highest level of initial inoculum (10%) and to the longest time ($4N$ = 40 days). Comparison of these curves with the three individual curves representing the same three r values in the graph at top right (row 1, column 3) shows the relative impact of reducing initial inoculum from 10% to 0.01%. Even the epidemic with the lowest r value progressed to 100% severity when initial inoculum levels were high and the time was long. On the other hand, progressing from bottom left (row 3, column 1) to top left (row 1, column 1), we see how reductions in initial inoculum affect disease progress when epidemics have less time to develop. Here even the highest r values (most aggressive pathogens, most susceptible hosts, and most favorable environments) are not effective because initial inoculum is low and time is short.

In terms of biological control of weeds, these graphs illustrate that even a slowly developing pathogen (low r value) could be useful as a biocontrol agent if sufficient time is available and inoculum can be distributed abundantly. Conversely, if the weed must be killed rapidly, a pathogen with a high r value should be selected to ensure that the epidemic could develop quickly. Alternatively, sufficient levels of inoculum should be applied at the start to ensure significantly higher initial levels of disease. One must first determine how the epidemic must progress for the desired control effects and then check the list of that weed's available pathogens. Once a pathogen is selected, on the basis of its epidemiologic capabilities and projected host impacts, it must be determined as to which of the various scenarios presented in Fig. 6 is most appropriate to the situation. For example, if an available pathogen has a high rate of reproduction r and a moderate amount of time t can be allowed for disease development, any of the curves of column 1 or 2 that go to completion sufficiently early to affect the host would define a satisfactory level of inoculation. If r is more moderate, the level of inoculation would have to be adjusted sufficiently higher to cause one of the bottom two curves of these graphs to reach early completion.

ON DEFINING CONTROL

Clearly, the process of formulating specific management goals must precede efforts to select from the available pathogens and the various epidemic curves that can be created. Before proceeding to discuss how specific relationships influence x_0, r, or t, some additional information is needed. Increase of disease includes both a change in disease intensity on a given host plant and a change in the number of host plants affected (area affected). The specific nature of that dual increase (spatial and temporal) closely corresponds to the quality and degree of weed control effected by the pathogen. Normally a disease does its damage by means of a chronic stress, which for a particular plant is related to a combination of the intensity of disease and the amount of time that intensity prevails (roughly related to the area under the disease progress curve—represented in the shaded areas of Fig. 4 or the areas beneath individual disease progress curves of Fig. 6). With crop pathogens, it has been shown that in general, the more severe a disease is and the longer that disease is present the more host damage that results (7,8). With some diseases, timing of disease pressure with critical host growth stages has an even greater influence (8). Timing is important for other reasons. The type of epidemic needed to prevent seed production by a weed (e.g., northern jointvetch in rice) may be one timed to peak later in the growing season than those needed to control weeds that compete with the growing plant for water, nutrients, or light. Other information such as crop growth habit and nature of the weed problem also help in making this determination.

Normally disease increases gradually, thus leading to a chronic stress of the host plant. However, heavy initial doses of some pathogens can cause shock reactions that, in turn, lead to quick death of the host. In these situations there is no need for further epidemic development. Such acute damage is well demonstrated in crop pathology. Although it may not be well documented for weeds, shock reactions are probably also common for weeds. Examples from crop pathology would include *Puccinia sorghi* on corn, bacterial blight of rice, and many of the soil-borne diseases of seedlings. The pathogens in these cases normally cause little or no chronic damage, yet when used in heavy amounts on young plants, they can quickly kill the host. If rapid death of the weed is preferred, management for acute host damage might be the chosen approach.

What constitutes economically justifiable levels of control has been debated among biocontrol workers, but no simple definitions suffice. For each new problem situation there may be a different answer, and for some situations there may be several answers. As an example, the speed of the epidemic needed to provide weed control in a pasture or undisturbed area may be much slower than the speed needed for effective control of the same weed competing with crops in a cultivated field. Control of the same weed growing both in crop lands and adjacent undisturbed areas might require a combined strategy—one part aimed at directly eliminating the weed competitor in the crop and the other at the seed sources of the weed. Adequate control would be defined in terms of combined success in both areas.

The formulation of goals may go beyond considerations for effectively killing weeds. Voids seldom exist in nature, and very rapid elimination of certain weeds may lead to replacement by more troublesome weeds. Competitive influences from various species which share ecosystem resources can be subtle yet decisive. Slight disturbances in a balance can lead to major population shifts. The target weed's survival may be sufficiently fragile that mild epidemics will control it. In some instances, mild epidemics may fit management goals better than explosive ones. For example, rapid death of large patches of *Chondrilla juncea* on roadsides and rangelands may lead to recolonization by *Carduus* thistle and other pioneer colonizers for which controls are not readily available or easily applied. A more gradual displacement of *Chondrilla* might allow the slower-spreading grasses and forage plants a better competitive advantage for the same space. Thus the primary objective of controlling this weed would be met more satisfactorily.

With this discussion, the shape of the desired epidemic should be fairly well elucidated. We can now proceed to create the epidemic.

SPECIFIC EFFECTS

Table 1 shows measurements of relative area under the disease severity curve from different scenarios of x_0, r, and t in Fig. 6. Areas of roughly the same magnitude will result in approximately the same amount of host stress. Determination of the necessary amount of stress for controlling a given weed requires quantitative evaluations. Within economic constraints, and noting the natural void exception mentioned above, one would normally attempt to maximize stress [i.e., use the most aggressive pathogen(s), distribute inoculum profusely, and apply the inoculum sufficiently early to provide maximum time for epidemic development]. Because biocontrol agents are host-selective, environmentally clean, and relatively cheap to

TABLE 1. Areas Under Disease Progress Curves from Fig. 6[a]

	Value of r per Unit	$t = 10$[c]	$t = 20$	$t = 40$
	0.1[d]	1.3	5.8	49.3
$x_0 = 0.01\%$[b]	0.4	10.5	301.5	2115.0
	0.9	201.4	1183.1	3180.0
	0.1	16.1	58.6	1301.0
$x_0 = 1\%$	0.4	90.4	807.9	2800.0
	0.9	439.0	1438.0	3435.0
	0.1	151.8	475.8	1810.1
$x_0 = 10\%$	0.4	424.2	1385.1	3380.0
	0.9	699.0	1700.0	3700.0

[a]The areas under disease progress curves give a relative measure of host damage. The larger the number, the larger the area under the curve, and the greater the potential impact because of greater intensity and/or duration of pathogen stress.

[b]x_0 = initial infection level.

[c]t = time available for disease development.

[d]The higher the r value, the more favorable the environment, more susceptible the host, or more aggressive the pathogen.

produce, the natural tendency to use a margin of safety by applying higher than optimum levels of inoculum can be encouraged without concern.

Table 1, in conjunction with Fig. 6, allows one to determine whether a given epidemic will accomplish the biocontrol task assigned. If it is determined from past experience that a disease pressure, for example, of 1000 severity units (time × intensity of disease) are necessary for control, one could use Table 1 and Fig. 6 to evaluate the various possibilities of x_0, r, and t for achieving control. The rate and timing of inoculation can be brought into consort with the expected range of r for that particular organism and climate so that 1000 or more severity units would be ensured. Determinations of the expected value of r and amount of disease sufficient to damage the host will be possible on the basis of past field experience or independent experiments designed for that purpose.

Figure 7 shows how x_0, r, and t, compare mathematically. This graph provides an additional perspective of the relationships described previously. The lower the level of initial inoculum, the greater the apparent delay in the epidemic at all r values. However, the relative impact of lowering x_0 is much greater with slow diseases (low r) than with faster ones (high r). The higher the r value for a given organism, the more a given addition of inoculum will promote the overall epidemic and thus the pathogen's impact on the weed.

Timely inoculations include several considerations. First, to give maximum time for epidemic development, inoculum should be applied as early as susceptible host tissue is available and weather is favorable. However, primary leaves may be shed naturally before reproduction from initial infections can occur. With obligate parasites, it may be necessary to ensure carry-over for secondary cycles in other ways. Second, whenever possible,

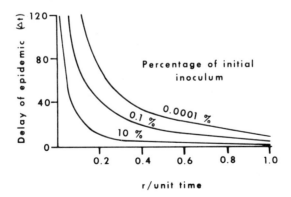

FIGURE 7. Interrelationship of x_0 and r in delaying epidemics. The time t required to obtain a given level of disease varies according to the level of initial inoculum x_0 and the rate of disease reproduction r.

inoculations should be timed to coincide with weather that is favorable for infection. Ideally, inoculations should be attempted prior to dew periods at which temperatures remain above 5° or 10°C. Application of inoculum is most effective in late afternoon and early evenings because clear nights often result in adequate dew, even during relatively dry periods.

Multiple inoculations can speed an epidemic. This may be necessary if sufficient moisture and suitable temperatures were not achieved following the first application of spores. Devices for the automatic, continuous release of inoculum have some advantages in this regard, but dispersal uniformity is equally important. To maximize dispersal, stationary devices that are used for inoculum dispersal should be placed high above the plant canopy and at multiple locations when possible to ensure good inoculum distribution.

Although the emphasis on the preceding discussions has been on fungal foliar pathogens, with minor adjustments the same principles hold for other classes of pathogens as well. Soil-borne pathogens are normally less easily disseminated (i.e., spatial isolation is more severe) than their foliar counterparts. Thus it is even more crucial that soil-borne inoculum be thoroughly disseminated. Otherwise, the same principles discussed apply to soil-borne pathogens.

COMPOUND EPIDEMIC

One note of further caution seems appropriate. For many of the same reasons that pathogens can be effective in weed biocontrol, organisms with life cycles shorter than that of the biocontrol agent could become a serious problem to the biocontrol effort. The earlier emphasis on the relative speed of reproductive cycles was partially meant to reinforce this point. It is possible that various hyperparasites (parasites of the biocontrol parasites) may be cultured inadvertently and disseminated along with biocontrol agents. "A flea upon the dog, a mite upon the flea, and so on, *ad infinitum.*" Such a mistake could nullify a biocontrol effort for years into the future. Our responsibility is not unlike the responsibility to avoid transporting dangerous weeds, insects, and pathogens into new crop areas when moving crops around the world. Preventive actions are necessary to ensure that the weed pathogen is not being produced and released in a parasitized condition.

CONCLUSIONS

To create an epidemic, one must begin with a pathogen that is capable of being virulent on the target weed and that is aggressive under local environmental conditions. Virulence and aggressiveness may require the use of exotic organisms, but in general that assumption is premature. Once a satisfactory organism has been selected, the biocontrol effort becomes an attempt to maximize the opportunity for reproduction of the pathogen.

Maximum reproduction is most easily facilitated by disseminating propagules as early in the cropping season as possible and as uniformly as possible across the host target area. Earliness and good distribution provide the opportunity for (1) maximizing the number of infection cycles, (2) taking better advantage of the generally moist and more moderate weather of spring, (3) minimizing the negative impact of isolation, and (4) maximizing the effective time for disease development on each member of the target weed population.

The quantity of inoculum is important from two perspectives: (1) more inoculum gives better overall distribution for earlier epidemics and (2) more inoculum gives a higher initial level of disease from which successive cycles of infection multiply. Any increase in initial inoculum is successively multiplied through new cycles of infection. Even a small increment in initial inoculum can eventually be multiplied into large increments of disease. As each cycle of infection may multiply inoculum several hundredfold, the application of inoculum one infection cycle earlier in the growing season can create an epidemic equivalent to that of several hundredfold increase in inoculum provided at a later date.

Regardless of how much and how well inoculum is disseminated, unless it is applied when an infection window is open or about to open, there will be little or no disease to show for the effort. The infection window includes compatibility of climate and susceptibility of host. The more favorable these are, the more effective the inoculum will be. Multiple applications of inoculum can greatly improve the opportunity for epidemics by (1) taking advantage of more of the infection windows, (2) increasing disease intensity, and (3) improving overall uniformity of dissemination.

Initial inoculum x_0, rate of reproduction r, and time t must be evaluated in consort. However, for the biocontrol setting, early seasonal application and effective distribution of inoculum deserve special recognition. Regardless of the pathogen chosen for use and the environment under which it must work, these two operations are the primary mechanisms by which successful epidemics can be managed.

REFERENCES

1. A. S. Crafts. 1975. *Modern Weed Control*. University of California Press, Berkeley, CA, 440 pp.

2. R. Daubenmire. 1968. *Plant Communities: A Textbook of Plant Synecology.* Harper and Row, New York, 300 pp.

3. M. J. Richardson. 1980. Agricultural Scientific Services, East Craigs, Craigs Road, Edinburgh EH12 8NJ, United Kingdom, personal communication.

4. R. B. Platt and J. F. Griffiths. 1964. *Environmental Measurement and Interpretation.* Reinhold, New York, 235 pp.

5. R. W. McMallen, L. M. Vaughan, and W. A. Perkins. 1968. *Wheat Stem Rust Epiphytotics.* Technical Report No. 152. Ft. Detrick, Frederick, MD, 55 pp.

6. J. E. van der Plank. 1963. *Plant Diseases: Epidemics and Control.* Academic Press, New York, 349 pp.

7. C. Populer. 1978. Changes in host susceptibility with time. Pp. 239–262 in: J. G. Horsfall and E. B. Cowling, eds., *Plant Disease: An Advanced Treatise.* Vol. 2. *How Disease Develops in Populations.* Academic Press, New York.

8. W. C. James. 1974. Assessment of plant diseases and losses. *Annu. Rev. Phytopathol.* 12, 27–48.

III

MICROBIAL WEED CONTROL AGENTS

9

MASS PRODUCTION OF MICROORGANISMS FOR BIOLOGICAL CONTROL

BRUCE W. CHURCHILL

Fermentation Research and Development, The Upjohn Company,
Kalamazoo, Michigan

During the last 10–15 years plant pathologists have reported a considerable volume of information on the use of plant pathogens for weed control. However, if widespread control of weeds with plant pathogens is to be accomplished, it is necessary to mass-produce certain microorganisms in the form of spores or other infectious inocula. The viable, infectious inocula must be produced rapidly in an inexpensive medium and be recoverable in an efficient and reproducible manner. The cost of producing microorganisms needed to treat large areas must be competitive with that of chemical pesticides.

The Upjohn Company has been involved for the last 7 years in the mass production of spores of *Colletotrichum gloeosporioides* f. sp. *aeschynomene* (CGA) for use as a biological herbicide. This fungus has been shown (1) to be effective in the control of northern jointvetch, *Aeschynomene virginica*, a serious weed in the rice and soybean fields in Arkansas. Much of the data presented in this chapter was obtained with CGA and two other fungal bioherbicides, *C. gloeosporioides* f. sp. *jussiaeae* (CGJ) and *C. malvarum* (CM). Limited attempts have also been made to mass-produce spores of several other microorganisms that attack specific weeds of fairly widespread distribution. The applications of modern fermentation technology methods and equipment used in these studies are discussed in detail. Although most studies have dealt almost exclusively with the production

139

of asexual spores, much of the methodology reported here could be applicable to the mass production of mycelial fragments and bacterial cells. Only after considerable experience will it be possible to determine whether other types of fruiting bodies can be mass-produced by these techniques.

STOCK CULTURE PREPARATION

The fermentation program starts with the establishment of methods of producing stable stock cultures. The stocks may be soil cultures, lyophilized cultures, agar slants, or spore suspensions. The stock cultures can be stored in a refrigerator or freezer or in the gas phase of a liquid nitrogen freezer. A good method of preserving cells, spores, or mycelia is to store them in tubes of sterilized soil. The inoculated tubes of soil are air-dried and stored at 4° or 25°C. Often a more usable storage is to place the cell or spore suspensions in ampuls that are sealed and frozen in the gas phase of a liquid nitrogen freezer. Cultures transferred from liquid nitrogen-frozen ampuls grow rapidly in nutrient media without the long lag phase that frequently occurs with soil and lyophilized cultures. Irrespective of the method of storage used, the working stocks must remain uniform over an extended period of time.

Some methods are used to test stock cultures to determine virulence and morphologic homogeneity. The cell suspensions are sprayed on the target weed and the degree of virulence observed. If less than expected virulence occurs, a new isolate is selected from an infected weed for the preparation of another stock culture. A micromanipulator is used to select a single spore from a spore suspension prepared from the infected weed tissue. The derivation of a stock culture from a single spore minimizes morphologic variability and the potential for loss of virulence. Preparation of stocks after several serial transfers of the culture will result in increased morphologic variability and the increased possibility of loss of virulence. It is necessary with some microorganisms to store the stock cultures in a medium that has the same osmotic concentration and usually the same nutrients as the growth medium to retain viability during prolonged storage.

SELECTION OF A SEED MEDIUM

The first step in the production of microorganisms is the screening of 20–30 agar media that are selected on the basis of past success in producing spores or other infective propagules of microorganisms. Most of these agar media are composed of readily available and inexpensive products that can supply nitrogen and carbon sources. The goal of mass production of microorganisms at an economical cost can be met by using crude agricultural products that are readily available in unlimited quantities. Generally, the medium used in liquid culture is a broth version of the agar medium that produces the most desirable growth. In the case of sporulating organisms, the choice of nutrients is often dependent on selecting those that give abundant sporulation. Protein sources such as soybean flour, corn steep liquor, distiller's solubles, brewer's yeast, autolyzed yeast, milk solids, cottonseed flour, linseed meal, corn protein, and a variety of fish meals are usually tested. The carbon sources tested include cornstarch, corn flour, glucose, hydrolyzed-corn-derived materials, glycerol, and sucrose. Occasionally, microorganisms that lack both amylase and sucrase have been found. Such organisms would restrict the choice of carbon sources in the culture medium.

In the case of CGA, Daniel et al. (1) reported good sporulation on lima bean agar and in a liquid medium composed of V-8 juice, sucrose, KNO_3, and mineral supplements. No special nutrition problems were encountered in the mass production of CGA since it grew and sporulated abundantly on several agar and liquid media tested. However, it was necessary to balance precisely the levels of nitrogen, carbon, and mineral supplement to give optimum spore yields. McCoy et al. (2) have reported the carbon-nitrogen combinations that gave the optimum production of the fungus *Hirsutella thompsonii*. The sporulation of *Colletotrichum lindemuthianum* was optimized by Mathur et al. (3) by balancing nutrients and selecting the best fermentation conditions. Latģe et al. and Latģe and Soper (4,5) have reported the production of zygospores of *Entomophthora virulenta* in a semidefined medium as well as in production-type media with crude protein and carbon sources.

Usually primary and secondary seed stages are used as a means of increasing spore titers rapidly without using massive initial inoculum in the fermentation. In addition, the seed stages greatly reduce the fermentation cycle of spore production. In this instance, the same medium can be used in both the primary and secondary seed vessels as well as in the production stage fermentation. Unlike antibiotic production, in which growth and antibiotic production phases require distinctly different media, in the case of microbial biological control agents, the production phase is only an extension of growth in the seed vessels.

STANDARDIZED INOCULUM

A standardized inoculum is the key to obtaining uniform spore yields. The inoculum is prepared in several shake flasks containing the seed medium, and the flasks are incubated under conditions optimal for abundant spore production. The inoculum for many future production runs can be prepared in the seed medium and stored in sealed ampuls. To obtain uniform growth from each ampul, the spore suspension must be continually agitated during the filling of the ampuls.

In determining the best conditions for storage of CGA in our laboratories, half of the prepared ampuls were purged with nitrogen before sealing and the others were sealed under ambient air. One ampul of each type was stored at 25°, 4°, −20°, or −172°C, with the latter in a liquid nitrogen vapor phase freezer. The ampuls were removed from storage at various times, and the spores were tested for germination.

In most cases the mycelia of fungi that have been tested remained viable after lyophilization, whereas the spores were unable to survive lyophilization. However, spores of CGA and other fungi tested survived prolonged freezing at −20°C, but several of them were not viable after storage for a short period in the liquid nitrogen freezer. None of the spores tested was viable in broth suspensions after prolonged storage at 4°C. The spores of CGA lost viability after a short period of storage in sealed ampuls in a −20°C self-defrosting freezer. Presumably the temperature fluctuations during the defrost cycles had a lethal effect on the spores, since spores stored in a freezer with a constant temperature of −20°C retained their viability for several months. Haynes et al. (6) have reviewed methods of preserving spores and vegetative microbial cultures. As they pointed out, the best method of culture maintenance would probably vary even with organisms of the same type, and thus, the method cannot be based entirely on a knowledge of related organisms.

SHAKE FLASK FERMENTATIONS

Shake flask fermentations are a necessity in the development of a process for the mass production of spores or other infective propagules of microorganisms. However, control of many parameters is difficult in shake flask fermentations. The lack of control over pH, temperature, agitation, and aeration makes the shake flask fermentation very difficult to use without considerable experience. A good degree of control can usually be achieved; for example, the pH may be controlled by using buffers, or the desired course of pH during the fermentation cycle may be obtained by using combinations of various levels of potassium nitrate, ammonium nitrate, and other utilizable ammonium salts. Temperature of the culture broth may be measured by using shake flasks containing thermometers in the side arms, and when the broth temperature exceeds the optimum for spore production as a result of active metabolism, the flasks may be shifted to a shaker room with lower ambient temperature. Aeration in shake flask fermentations can be markedly increased by reducing the concentration of the medium or the volume of the medium per flask, using various types of baffles (the stippled being the most effective), and/ or increasing the speed of the rotary shaker. The converse of higher aeration methods is used to reduce aeration, and in addition, a restriction of the size of the neck opening reduces the amount of air diffusing into the flask.

Several of these controls over parameters were tested in the shake flask fermentation studies with CGA, CGJ, and CM. A very high level of aeration was required to give optimum spore production with CGA and CM, whereas a low level of aeration was essential for heavy spore production with CGJ. The required aeration level was obtained by using a reduced volume of a reduced strength medium for CM and a reduced volume of a full-strength medium for CGA. All three fungi produced abundant mycelial growth and very low spore numbers when the aeration levels were not optimum for spore production. (For the effects of aeration levels on spore production in tank fermentations, see the section on scaling up from shake flasks to production tanks.) A large number of shake flask fermentations were run to determine the optimum conditions for maximum spore production. A statistical, factorial 2^5 design was used that gave for each group of 32 shake flasks an indication of the interactions obtained with all possible combinations of two levels of five variables. Using this method, the beneficial changes in mineral supplement, the effects of various carbon and nitrogen sources, the carbon:nitrogen ratio, and the need for vitamins and growth factors for maximum spore production by CGA, CGJ, and CM were determined. With the three fungi the carbon source was critical, but not the levels or types of nitrogen or the carbon:nitrogen ratio. The carbon:nitrogen ratio was reported to be critical for the production of *H. thompsonii* (2). Also, CGA, CGJ, and CM did not require vitamins and growth factors for optimum sporulation, unlike those required by *C. lindemuthianum* in studies by Mathur et al. (3).

The duration of the various growth cycles, namely, the primary and secondary seeds and the submerged production cycle, were found to be critical with some of the fungi studied. The production cycle with CGA was terminated at about the same time in each run to obtain good germination of the spores at harvest. Extension by 2 additional days of either the shake flask seed or the production cycle resulted in a substantial reduction in the number of viable spores even if additional nutrients were added. Microscopic examination of the spores showed that many had become nonrefractile, and in most studies a loss of refractility was associated with failure of spores to germinate.

METHODS OF MAINTAINING STERILITY

The most challenging aspect of the mass production of microorganisms for biological weed control is the maintenance of sterility. The presence of contaminating bacteria is of primary concern in the production of microbial herbicides since bacterial contaminants could affect the efficacy of inocula. The initial phase of sterility control is the preparation and storage of contamination-free primary working inoculum. The preparation of this inoculum requires that many sterility checks be run with pH-sensitive dyes such as thioglycollate or phenol red, nutrient agar media, and by microscopic examinations. The bacterial contamination problem is much more acute in microbial herbicide production than in submerged antibiotic fermentations. In the latter, low levels of contaminants are destroyed by the antibiotic produced. The microbial herbicides studied have not been shown to produce antibacterial compounds, and the nutrients and pH of these fermentations are usually ideal for bacterial growth.

Particular precautions should be taken to maintain sterility in fermentations. Fermentors with which difficulties are encountered in maintaining sterility should not be used for mass production of these microorganisms. Likewise, a fermentor should not be used until contamination problems have been eliminated. Ideally, a fermentor should be used only for the mass production of biocontrol microorganisms, but that goal is not practical unless there is a need for continuous production of those agents. Some special equipment such as mechanical seals around the agitators, although very expensive, might further reduce the problem of contamination.

It is preferable to make all transfers through steam-sterilized lines during the various stages of the fermentation cycle to minimize exposure of the cultures to airborne contaminants. In our fermentations, a so-called closed-system primary seed is grown on a rotary shaker. The closed-system vessel is designed so that the medium can be inoculated with a sterile syringe through a sterile membrane in the side arm of the vessel. The primary seed at harvest can be forced into the seed tank, through a steam sterilized line, without opening the closed-system vessel. The seed tanks, fermentors, and auxiliary piping are sterilized with pressurized steam. The many ports used for additions and measuring several parameters during the fermentation cycle further increase the chances of contaminants entering the fermentor. A positive pressure is maintained in the tanks, usually at 5–10 psi, to ensure that air will be passing outward even if minute leaks are present. Sterile dry air is used to supply oxygen to the seed tanks and fermentors. Dry air is sterilized by passing it through banks of deep-bed fiberglass or cartridge-type membrane filters. The latter are supplied in a variety of sizes by Nucleopore, Pall, or Millipore. Very large scale fermentations may require over 100,000 liters/minute of air, which is one of the problems involved in sterilizing air for a large production plant. Other problems and methods involved in sterilizing air streams and fermentation equipment have been covered in a review by Parker (7).

The method of growing microorganisms on nonliquid media in roller drums or stationary trays as an alternative to submerged liquid culture would still present some serious sterility problems. For example, the problems associated with delivering sterile dry air and of the initial sterilization of the vessels would be similar to those encountered with deep-tank fermentations. Hesseltine (8) has reviewed the use of solid or semisolid substrates in roller drums to produce dry spores of *Aspergillus oryzae* that retained their viability even after grinding. Some microorganisms produce more spores in semisolid growth vessels, but sampling and sterility problems are often very acute. The major advantages of semisolid or solid substrates in mass production of fungal spores is discussed in the section on harvesting and drying of fungal spores.

SCALING UP FROM SHAKE FLASKS TO PRODUCTION TANKS

Some sophisticated facilities are now available to scale up fermentations. Many small microbial fermentors of 9–20-liter operating volume are available for preliminary scale-up studies. These fermentors are particularly useful since they can provide a wide range of pH, agitator speeds, impeller designs, aeration rates, choice of incoming gases, variations in baffling and backpressures, and temperatures. Many large-scale fermentors with geometric design similar to the smaller microbial fermentors are available. The consistency of design simplifies scale-up studies. Pilot plant fermentors with 500-liter operating volume were used to produce CGA and CGJ spores for field trials and to determine scale-up factors. Computers can greatly aid in scaling up the mass production of spores by automatically monitoring and recording a number of parameters such as pH, airflow, carbon dioxide levels, oxygen uptake, and dissolved oxygen. Computers can also direct the analysis and recording of gas profiles continuously during the course of many parallel fermentations.

The ability to analyze and record the level of gases in the effluent streams from both the seed tank and the fermentor is often essential. Arnold and Steel (9) have suggested that the respiration rate of an organism is dependent on the type of substrate and the state of physiological development of the culture. Moreover, the critical level of oxygen tension required to obtain the maximum rate of oxygen utilization with some microorganisms may be comparatively high, and the diffusion of oxygen molecules through cell protoplasm may have a much greater role in controlling the respiratory rate as cells become larger (9). Also Corman et al. (10) have reported that the oxygen absorption rates in various types of equipment can be a critical factor in the success of scale-up of a fermentation.

Computer plots of dissolved oxygen, oxygen uptake, and CO_2 produced in seed tanks and fermentors of CGA and CGJ at high and low rates of agitation and aeration enabled a study of the relationship of these parameters to sporulation of the two fungi. When the agitation and aeration were sufficiently high in the seed tanks, the level of dissolved oxygen was not appreciably reduced, even at the peak of oxygen uptake by CGA (Fig. 1a). Even with drastic reduction in agitation and aeration of CGA seed tanks (Fig. 1b), spore titers were normal, no mycelia were present at harvest, and a moderate rise in oxygen uptake and CO_2 production resulted. The dissolved oxygen was reduced to about 75% of saturation at harvest. When CGA seed cultures grown at reduced agitation and aeration were used as inoculum in a high-agitation and aeration fermentation, these gave normal spore production for 24 hours. However, the spore titer at harvest was substantially lower than at 24 hours, and examination of the culture growth showed that it was a mass of mycelia. This is strong evidence that the type of growth in the production fermentors was preconditioned by the levels of agitation and aeration in the seed tanks. Thus spores produced under high-dissolved oxygen (high agitation and aeration) in the seed tanks produced fission spores in subsequent fermentations, whereas spores produced in seed tanks under low-dissolved oxygen produced only mycelial growth in subsequent fermentations. The possibility that the high aeration levels sweep CO_2 and/or other volatiles from the seed tanks and fermentors is unlikely to be the cause of production of spores at high aeration levels because identical results are obtained in highly agitated and aerated shake flasks in which gas exchange is slowed by diffusion through the cover. Generally, spores are produced as a result of conditions unfavorable for vegetative growth such as the depletion of nutrients, temperature extremes, and unsuitable pH. In the case of CGA, spores were considered to have been produced in seed tanks and

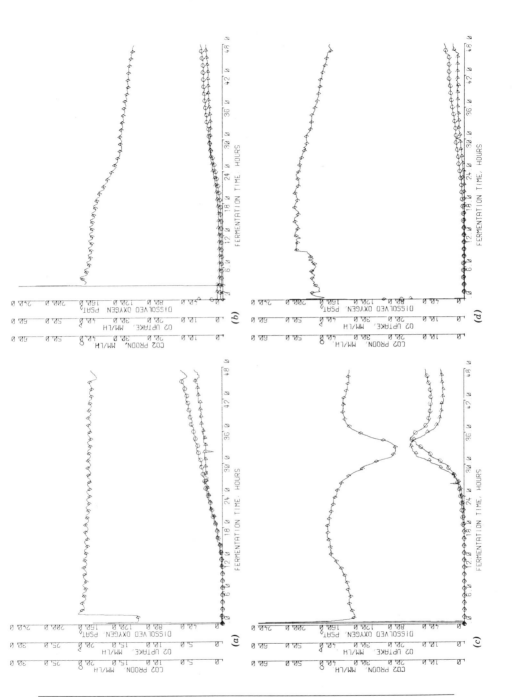

FIGURE 1. Relationship of the rates of agitation and aeration to the dissolved oxygen levels of cultures and the respiration of CGA and CGJ in seed tanks: (*a*) CGA tank with high agitation and aeration; (*b*) CGA tank with low agitation and aeration; (*c*) CGJ tank with low agitation and aeration; (*d*) CGJ tank with high agitation and aeration. $\Diamond$ —Dissolved O_2 as a percentage of saturation level, $\triangleright$ —O_2 uptake as mM/(liter·hour); and $\bigcirc$ —CO_2 produced as mM/(liter·hour).

fermentors as a result of very high agitation and aeration, and consequently the unfavorably high oxygen tension.

Even though CGA and CGJ are very closely related fungi, their responses to oxygen tension in the medium were vastly different. For instance, CGJ produced masses of spores only if low agitation and aeration were used continuously in both the seed and production cycles. Spore production with CGJ ceased in any stage when the agitation and aeration were too high, resulting in a mass production of mycelia. Figure 1c represents the levels of gases in a CGJ seed tank with low rates of agitation and aeration. Spore production in this tank was initiated when dissolved oxygen was reduced to about 80% of saturation during the first few hours of the seed cycle. Additionally, a sharp rise in oxygen uptake occurred during the period of growth and maturation of fission spores to the much larger slipper-shaped spores that are identical in appearance to conidia. This rise in oxygen occurred 33–42 hours after inoculation (Fig. 1c).

Figure 1d represents a CGJ seed tank with high agitation and aeration levels. The levels of dissolved oxygen recorded under these conditions were not noticeably reduced by oxygen uptake. The use of this seed as inoculum in a high agitation and aeration fermentor resulted in heavy mycelial growth. In summary, CGJ produced abundant spores under low agitation and aeration in shake flasks, seed tanks, and fermentors, whereas the opposite was true with CGA.

Data of this type obtained from pilot plant fermentations are useful in scaling up of spore production processes in tanks of up to 150,000-liter capacity. Occasionally, modifications will be needed in the agitation and aeration rates or in the composition of the medium to achieve the desired dissolved oxygen levels in very deep fermentation tanks.

INNOVATIONS IN THE MODERN PRODUCTION PLANT

The increased efficiency of modern industrial fermentation plants has been due largely to automation and sophisticated instrumentation. Automation has also substantially reduced labor costs. Automatic weighing and measuring of ingredients are now commonplace. The ever greater use of bulk storage silos, which has led to the elimination of bagged ingredients, has also been a key to automation and lowering of ingredient costs. The size of fermentors in production has increased from 15,000–30,000 liters to 150,000–200,000 liters or greater during the last 15–20 years. The larger batch fermentations require about the same work-hours of labor as the much smaller fermentations. The replacement of the reciprocal compressors with high speed turbines has made it possible to supply sufficient aeration to production tanks at a reduced cost. The use of centrally located instrument panels on which numerous parameters are simultaneously measured for many fermentations have further reduced labor costs. The use of automatic pH adjustment through feeds of alkali or acid, adjustment of airflow rate to match the fermentation demand, temperature shifts at predesignated times, additions of antifoam as needed, and the continuous, slow additions of a variety of nutrients or precursors are all possible. The end result of these many improvements is the vastly efficient microbial fermentations. These improvements should ensure versatility and economical operation in the mass production of bioherbicides at a cost that is competitive with that of chemical herbicides.

SPORE PRODUCTION BY FUNGAL PLANT PATHOGENS

Templeton (unpublished data) observed that CGA formed conidia in acervuli on the host plant northern jointvetch as well as on the surface of agar in petri plates. Daniel et al. (11) reported the production of abundant spores of CGA in shake flask cultures and magnetically stirred small fermentors; this finding suggested that deep tank fermentations with gentle agitation might yield spores. In the conventional agitator-stirred fermentors a very high shear occurs at the periphery of the impeller. This high shear could rip the mycelia and thus greatly reduce the production of conidia that normally appear on long branching conidiophores. Deindoerfer and Gaden (12) found that shear (yield stress) increases with the age of culture and the concentration of the organism. Therefore, the presence of a high concentration of CGA spores and the virtual absence of mycelia at harvest in the high-speed agitated fermentation were surprising and led us to a time course study. Quite unexpectedly, the results of this study indicated that the spores in the fermentor were produced in various ways. The majority of the spores, which at maturity were identical in appearance, size, and shape to the conidia produced in acervuli, were produced by binary fission (Fig. 2a). The fission spores were small and round immediately after fission. In a few hours the fission spores became slipper-shaped and enlarged to the size of typical conidia. In addition, spores were produced on conidiophores on very short branching hyphae (Fig. 2b); the shortness of hyphae resulted from continuous stirring as well as shear of the agitation. Blastospores (Fig. 2c) also were produced, from either a slipper-shaped spore (top) or from a fission spore

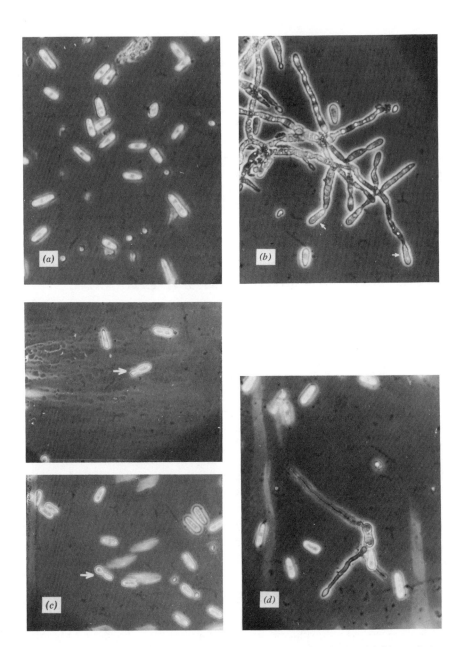

FIGURE 2. Types of spores produced by CGA in deep-tank cultures: (*a*) binary fission spores; (*b*) conidia (arrows) on short, highly branched hyphae; (*c*) two stages in blastospore formation; (*d*) a binary fission spore producing a secondary conidium at the end of a germ tube.

150

(bottom), fairly frequently by CGA from the midpoint of the fermentation cycle. The production of blastospores in deep-tank cultures is not unusual. Dulmage and Rhodes (13) have reported that many entomopathogenic fungi produce blastospores rather than true conidia in submerged cultures. They also reported that the blastospores produced in submerged cultures were very difficult to preserve and easily lost their viability and, therefore, that conidia are more useful in microbial insecticides (13). Frequently, toward the end of the fermentation cycle the CGA hyphae underwent fragmentation and matured into arthrospores. Secondary conidia formed on germinating fission spores were also seen (Fig. 2d). On several occasions characteristic acervuli-like structures were also recovered from the walls of fermenters above the liquid level. It is estimated that only about 8–10% of the CGA spores produced were conidia. Most others, estimated at 80–85%, were identified as being derived from fissions. The blastospores and arthrospores total less than 5% of the spores produced.

Like CGA, the winged waterprimrose pathogen CGJ (14) produced a variety of spores in fermentation cultures. The binary fission spores of CGJ were produced in about the same proportions as in CGA cultures. However, the blastospores were much more numerous, and true conidia and arthrospores were fewer with CGJ than with CGA.

Colletotrichum malvarum, the pathogen of prickly sida, has also been observed to produce fission spores as well as conidia. The other fungal organisms that have been studied did not appear to produce fission spores or blastospores, although they produced conidia. Undoubtedly, the methods of spore production will vary markedly from one organism to another. For example, *Alternaria macrospora*, a pathogen of spurred anoda (15), failed to produce spores in liquid cultures without exposure to periods of alternating light and darkness.

Particularly fragile organisms should be handled under mild conditions such as offered by the airlift fermentor described by Lundgren and Russell (16), in fermentors with slow-rpm impellers and large blades, or even in surface cultures. One might predict that the bioengineers will be called on to alter or design new types of fermentor to meet the special needs for mass production of spores of some fungal pathogens. How soon these new fermentors are designed will depend on the size of the potential market for new microbial biocontrol agents. The concept of mixing many pathogens to control all major weeds in a field could greatly improve the eventual sale of bioherbicides. This concept was described by Boyette et al. (14). A mixture of dried spores of CGA and CGJ was rehydrated with cold water and then sprayed on rice fields in Arkansas with subsequent success in controlling both northern jointvetch and winged waterprimrose.

HARVESTING AND DRYING OF FUNGAL SPORES

It is obvious that each fungus should be allowed to reach a certain age in culture before it will yield the desired spore titer. Therefore, the length of the fermentation cycle of each fungus must be determined to achieve consistently uniform spore titer, spore maturity, and spore viability.

The separation of spores from the aqueous broth can be accomplished by different methods. The most common methods are centrifugation in a DeLaval- or Sharples-type centrifuge, filtration through a plate and frame filter, and removal of spores with a rotary vacuum filter (e.g., an Eimco) precoated with a filter aid. It should be apparent that if fungal biocontrol agents are to become marketable products, their spores or other infective propagules should be held in a stable and viable state. The spores that are filtered or otherwise separated from the culture fluid must be dried rapidly but gently. The particular set of conditions for drying will vary with each organism. The common methods used include freeze-drying, air-drying, spray-drying, and vacuum-tumbler-drying as well as various combinations of these methods. One modification is the use of a pelletizer before the spores are air-dried. A method that is sufficiently mild for the spores of one organism may totally destroy the viability of another. The general rule in drying spores seems to be the selection of a method that assures good viability of dried spores at the most economical cost. The statement by Kenney et al. (17), "under submerged culture conditions, many fungi do not produce a stable spore form" may be true only until suitable conditions are found to process and dry fungal spores in a viable form. Some attempts have been made to dry fungi that do not produce spores. If mycelia are viable after drying and milling, this alternative is probably preferable to the growing of cultures in semisolid or solid fermentations.

Roller drums and tray cultures have been used for the mass production of some fungi that do not produce spores in deep-tank fermentations. The solid substrate fermentation has also been used for fungal pathogens, the mycelia of which do not remain viable after drying and grinding, or where no method of stabilizing the spores has been found. The method of producing fungal spores on bran has resulted in very high levels of spore production with some organisms. A roller drum fermentation method, similar to the one described by Hesseltine (8), has been used by Dulmage and Rhodes (13) for the production of several fungal entomopathogens. A disadvantage of solid or semisolid fermentations, in addition to the extreme difficulties in maintaining sterility and the lack of controls on fermentation conditions, is the problem of recovering the spores from the substrate. The spores recovered must still be dried to maintain an adequate shelf life. An air-drying cycle in roller drums may be a distinct advantage over deep-tank fermentations if the spores are able to withstand grinding at harvest. The fact that CGA, CGJ, and CM spores are infective on the particular host weed regardless of the method of spore production is probably not a general rule.

GERMINATION OF DRIED SPORES

Samples of dried spores must be germinated to assay for viable spores, check for contamination, and determine the shelf life of the product. Although a pure culture may be free of contamination at harvest, contaminants may enter the spore preparation during its concentration and drying. Generally, the occasional organism present in spore preparations has not been observed to multiply during the rapid drying process. However, contaminants may increase in numbers during the incubation period of the germination assay. Luckily, such occasional contaminants have not affected the germination of CGA, CGJ, or CM.

The early assays of fresh spores and spore cakes of CGA posed no problem in either sampling or obtaining uniform germination. The fresh spores were filtered and washed three times, and the spore suspension was diluted to 1–2 million spores/ml. The spore suspensions were stirred to break up spore clumps, drops of the spore suspension were then placed on water agar in petri plates, and the plates were incubated at 28°C for 8–12 hours. The germinated spores had germ tubes at least as long as the length of the spore. Typical suspensions gave 90–100% spore germination.

When dried spores of CGA were assayed by testing their metabolic activity with the oxidation-reduction dye 2,3,5-triphenyltetrazolium chloride (18), a substantially higher percentage of spores were found to reduce the dye than those germinated on water agar. Since fresh spores did not have special nutrient requirements for germination, the reduced germination of dried spores should be due to either dormancy, drying, or a requirement for very slow hydration. Various methods were tried to induce a higher percentage of dried CGA spores to germinate. Different carbohydrates were added to the spores as described by Gottlieb (19) and Lin (20) without any increase in the percent germination. Irradiation of the spores with ultraviolet at various wavelengths, which improved the germination of fresh spores of *Colletotrichum phomoides* (21), also failed with dried CGA spores. A large number of compounds, reported by Smith and Berry (22) to effect fungal spore germination, were also screened without success.

An improved method of germination of dried CGA spore preparations was devised, as follows. High concentrations of spores were suspended and ground in cold water, counted with a hemacytometer, and refrigerated for at least 2 hours before dilution-plating on water agar at 28°C. Soaking in the refrigerator allowed the spores to hydrate without germination. This method helped to increase germination of most of the dried spore preparations by 20–30%. The dried spores, presoaked in cold water for 2 hours, diluted, and sprayed at the rate of 187×10^9 spores/ha (23) on northern jointvetch-infested rice fields confirmed the efficacy of the improved germination method. The infection levels of fresh and dried spores germinated by this procedure were equal, as was the percentage of weeds killed. It is theorized that the dried spores must be hydrated slowly and fully prior to exposure to conditions favorable for germination. Incomplete hydration in the process of germination would have a lethal effect on the spores. Dried spores of some microorganisms might require a still slower rate of hydration. Very slow hydration can be achieved by controlling humidity or by reducing the osmotic pressure on the spores by controlling solutes in the external solution. In the case of CGA, free water and not just 100% humidity was required for spore germination, and this is also true in the case of the conidia of *Sclerotinia fructicola* (24). During the last 5 years several dried spore preparations of CGA have been sprayed on rice and soybean fields in Arkansas. The germination assays of these preparations have ranged from about 20 to 76%, and at these levels, all the dried preparations tested have provided good control of northern jointvetch. This success suggests that the spores produced in deep tank fermentations have always been infective. Spores of a *Colletotrichum* species grown at a higher than optimum temperature and pH have been reported by Ling and Yu (25) to be resistant to thermal death. If so, this may allow the production of spores that can withstand higher temperatures during culture, storage, and drying.

CONCLUSIONS

Marketable quantities of viable dried spores of CGA and CGJ that provided good biocontrol of the respective host weeds northern jointvetch and winged waterprimrose have been produced. Presently available methods of fermentation technology were used to produce these spores. Both CGA and CGJ have adapted to fermentation conditions by producing types of spores, most of which were different from the conidia produced on the host weeds. Despite this difference, the majority of the cultured spores were pathogenic on the host weeds, and the problem weeds could be controlled with aerial spraying of 187×10^9 dried viable spores/ha. The industry can supply this level of viable spores at a price that would appear to be competitive with those of the chemical herbicides that are in use. A mixture of dried viable spores of CGA and CGJ has been prepared and applied in one aerial application, resulting in the control of both target weeds in a rice field in Arkansas.

The fermentation expertise available in the pharmaceutical industry offers the best hope of reaching the goal of mass production of microorganisms for biocontrol of weeds. The limited experience discussed in this chapter would suggest that plant pathologists, microbiologists, and engineers must cooperate fully to solve the problems involved in the mass production of microbial herbicides.

Combinations of several factors can cause inhibition of sporulation in submerged fermentations (26). Smith and Anderson (27) reported that the reduction or absence of assimilable nitrogen coupled with available carbon generally induces sporulation. Fungal differentiation studies require that mycelia be produced under controlled conditions of aeration, temperature, agitation, pH, incremental feeding, and medium replacement (28).

We cannot hazard a guess as to the percentage of plant pathogens, of those that will be discovered, that can be mass-produced in deep-tank fermentations. Nor can we, with the limited studies that have been done, predict what other reproductive structures can be produced, with or without modifications to the currently available fermentors. Therefore, it is reasonable to "expect the unexpected" in microbial fermentations.

REFERENCES

1. J. T. Daniel, G. E. Templeton, R. J. Smith, Jr., and W. T. Fox. 1973. Biological control of northern jointvetch in rice with an endemic fungal disease. *Weed Sci.* 21, 303–307.

2. C. W. McCoy, T. L. Couch, and R. Weatherwax. 1978. A simplified medium for the production of *Hirsutella thompsonii. J. Invertebr. Pathol.* 31, 137–139.

3. R. S. Mathur, H. L. Barnett, and V. G. Lilly. 1950. Sporulation of *Colletotrichum lindemuthianum* in culture. *Phytopathology* 40, 104–114.

4. J. P. Latǵe, G. Remaudiere, R. S. Soper, C. D. Madore, and M. Diaquin. 1978. Growth and sporulation of *Entomophthora virulenta* on semidefined media in liquid culture. *J. Invertebr. Pathol.* 31, 225–233.

5. J. P. Latǵe and R. S. Soper. 1977. Media suitable for industrial production of *Entomophthora virulenta* zygospores. *Biotechnol. Bioeng.* 19, 1269–1284.

6. W. C. Haynes, L. J. Wickerham, and C. W. Hesseltine. 1955. Maintenance of cultures of industrially important microorganisms. *Appl. Microbiol.* 3, 361–368.

7. A. Parker. 1953. Sterilization of equipment, air and media. Pp. 97–121 in: R. Steel, ed., *Biochemical Engineering.* Heywood, London.

8. C. W. Hesseltine. 1977. Solid state fermentation. *Process Biochem.* **12**, 24–27.

9. B. H. Arnold and R. Steel. 1953. Oxygen supply and demand in aerobic fermentations. Pp. 149–181 in: R. Steel, ed., *Biochemical Engineering.* Heywood, London.

10. J. Corman, H. M. Tsuchiya, H. J. Koepsell, R. G. Benedict, S. E. Kelley, V. H. Feger, R. G. Dworschack, and R. W. Jackson. 1957. Oxygen adsorption rates in laboratory and pilot plant equipment. *Appl. Microbiol.* **5**, 313–318.

11. J. T. Daniel, G. E. Templeton, and R. J. Smith, Jr. 1974. Control of *Aeschynomeme* sp. with *Colletotrichum gloeosporioides* (Penz.) Sacc. f. sp. *aeschynomeme.* U.S. Patent No. 3,849,104.

12. F. H. Deindoerfer and E. L. Gaden, Jr. 1955. Effects of liquid physical properties on oxygen transfer in penicillin fermentation. *Appl. Microbiol.* **3**, 253–257.

13. H. T. Dulmage and R. H. Rhodes. 1971. Production of pathogens in artificial media. Pp. 507–538 in: H. D. Burges and N. W. Hussey, eds., *Microbial Control of Insects and Mites.* Academic Press, London.

14. C. D. Boyette, G. E. Templeton, and R. J. Smith, Jr. 1979. Control of winged water-primrose (*Jussiaea decurrens*) and northern jointvetch (*Aeschynomeme virginica*) with fungal pathogens. *Weed Sci.* **27**, 497–501.

15. H. D. Ohr, F. G. Pollack, and B. F. Ingber. 1977. The occurrence of *Alternaria macrospora* on *Anoda cristata* in Mississippi. *Plant Dis. Rep.* **61**, 208–209.

16. D. G. Lundgren and R. T. Russell. 1956. An air lift laboratory fermentor. *Appl. Microbiol.* **4**, 31–33.

17. D. S. Kenney, K. E. Conway, and W. H. Ridings. 1979. Mycoherbicides—potential for commercialization. *Dev. Ind. Microbiol.* **20**, 123–130.

18. R. B. Fred and S. G. Knight. 1949. The reduction of 2,3,5-triphenyltetrazolium chloride by *Penicillium chrysogenum*. *Science* **109**, 169–170.

19. D. Gottlieb. 1950. The physiology of spore germination of fungi. *Bot. Rev.* **16**, 229–257.

20. C. K. Lin. 1945. Nutrient requirements in the germination of the conidia of *Glomerella cingulata. Am. J. Bot.* **32**, 296–298.

21. A. H. Hutchinson and M. R. Ashton. 1930. The effect of radiant energy on growth and sporulation in *Colletotrichum phomoides. Can. J. Res.* **3**, 187–199.

22. J. E. Smith and D. R. Berry. 1974. *An Introduction to Biochemistry of Fungal Development.* Academic Press, London, 326 pp.

23. G. E. Templeton, D. O. TeBeest, and R. J. Smith, Jr. 1978. Development of an endemic fungal pathogen as a mycoherbicide for biocontrol of northern jointvetch in rice. Pp. 214–216 in: T. E. Freeman, ed., *Proceedings of the Fourth International Symposium on Biological Control of Weeds.* University of Florida, Gainesville, FL.

24. C. N. Clayton. 1942. The germination of fungous spores in relation to controlled humidity. *Phytopathology* **32**, 921–943.

25. L. Ling and E. H. Yu. 1941. Thermal death point of fungi in relation to growing conditions. *Phytopathology* **31**, 264–270.

26. G. Turian and D. E. Bianchi. 1972. Conidiation in *Neurospora. Bot. Rev.* **38**, 119–154.

27. J. E. Smith and J. G. Anderson. 1973. Differentiation in the aspergilli. Pp. 295–337 in: J. M. Ashworth and J. E. Smith, eds., *Microbial Differentiation. Twenty-third Symposium Society of General Microbiology.* Cambridge University Press, London.

28. A. T. Bull and M. E. Bushell. 1976. Environmental control of fungal growth. Pp. 1–31 in: J. E. Smith and D. R. Berry, eds., *The Filamentous Fungi.* Vol. 2. Wiley, New York.

10

COMMERCIALIZATION OF MICROBIAL BIOLOGICAL CONTROL AGENTS

R. C. BOWERS

Agricultural Division, The Upjohn Company,
Kalamazoo, Michigan

Biological control agents will probably assume an increasingly prominent role in pest management as we look toward new, effective, and environmentally responsible technologies to supplement chemical pest control in agriculture. In the future, control of many pests through the use of microorganisms may become the technology of choice. Scientific interest and redirection of resources toward the discovery of biological agents for control of agricultural pests have increased significantly in recent years.

The term "biological control agent" has different meanings for different interest groups. The U.S. Environmental Protection Agency (EPA) has coined the term "biorational pesticides" (see Chapter 11 for more on biorational pesticides) to describe agents that have a biological rationale for their use. Further, the EPA has divided biorationals into biochemical and microbial pest control agents. Biochemicals include compounds such as insect pheromones, growth regulators, and hormones (insect and plant), whereas microbial pest control agents include fungi, bacteria, viruses, and protozoa (1).

The EPA has specifically excluded macrobials such as predacious insects, parasites, and nematodes from the definition of biorational pesticides. The term "control" is also open to subjective judgment. The EPA has defined control as a 70–100% reduction in a pest population and has used suppression, aids in control, or other such terms for less than a 70% reduction in the target population. The EPA definitions may be open to

157

scientific debate. However, in this chapter the term "microbial pesticides" refers to fungi, bacteria, viruses, and protozoa, and the term "control" refers to a 70–100% reduction in a pest population (2).

Currently, there are 10 microbial pest control agents registered by the EPA. These include five bacteria (*Bacillus thuringiensis* for control of lepidopterous larvae, *B. popilliae* and *B. lentimorbus* for control of Japanese beetle larvae, *B. thuringiensis* var. *aizawai* for control of wax moth, and *Agrobacterium radiobacter* for control of crown gall); three viruses [the polyhedral inclusion bodies of *Heliothis* nuclear polyhedrosis virus (NPV), Douglas fir tussock moth NPV, and gypsy moth NPV]; one protozoan (*Nosema locustae* for control of grasshoppers); and one fungus (*Phytophthora palmivora* for control of milkweed vine in citrus in Florida). When this list of 10 is compared with the list of approximately 1500 registered chemical pesticides, it is easy to see that biological control agents have not been a major component of modern agricultural pest management programs.

The greatest emphasis in biological control has been directed at control of insects, mites, and weeds using the classical approach (3), which relies on the importation of self-sustaining and self-disseminating biological agents to control exotic or native pests. This approach has produced some of the most spectacular results, one of the most notable of which is the control of cottony cushion scale on citrus in California through the introduction of the vedalia beetle in 1899 (4). In contrast to the classical approach to pest control, the microbial pesticide approach involves the deliberate manipulation of biocontrol agents by artificial propagation and distribution. The most commercially successful microbial pesticide is *B. thuringiensis* (4). Hall et al. (4) have reported that the worldwide rate of "complete success" in classical biological control of arthropods is only 0.16 (n = 602) and the rate of "any success," which includes both complete and partial success, is 0.58. Further, it is estimated (5) that microbial pesticides constitute less than 1% of total pesticide sales in the world. Hence this history of achievement is not an obvious incentive for industry to commercialize either macrobial or microbial control agents.

Several studies have attempted to project the role that biological control agents will play in pest management strategies of the future (3, 5–7). For the most part, these studies have projected modest gains in the use of biological controls. These studies also indicate that if biological agents are to show any significant degree of success, the following must occur: (1) integrated pest management systems must be enhanced through an increase in the knowledge base, improvement in communication, environmental monitoring systems, and pest management training programs; this can be accomplished by overcoming grower skepticism through better coordination of programs and relaxation of cosmetic (aesthetic) standards; (2) government support of research and development on biological control must be continued; and (3) stringent regulation of chemical pesticides must be continued.

ROLE OF INDUSTRY IN COMMERCIALIZATION OF MICROBIAL BIOLOGICAL CONTROL AGENTS

In recognition of the growth potential for the use of biological control agents, industry has undertaken a modest but positive effort toward commercializing microbial pesticides. The Upjohn Company, for example, has collaborated with the University of Arkansas and the U.S. Department of Agriculture (USDA) in developing *Colletotrichum gloeosporioides* f. sp. *aeschynomene* (CGA) as a microbial herbicide for control of the leguminous weed *Aeschynomene virginica* (northern jointvetch). The weed is limited in geographic distribution to a relatively small hectarage in Arkansas and reduces yields and quality of rice (8). Further discussions on this pathogen and its weed host can be found in Chapters 3, 9, and 12.

The Upjohn Company recognizes four phases to commercial development of microbial control agents: (1) discovery; (2) lead development; (3) product development; and (4) marketing. The first three phases are discussed in this chapter; further views on commercialization of biological control agents can be found in other articles (9–14).

DISCOVERY

ROLE OF INDUSTRIAL VERSUS NONINDUSTRIAL SECTORS

Traditionally, private industry has discovered chemical pesticides by using screening systems devised to evaluate biological activity of synthetic compounds. Once pesticidal activity is found, a comprehensive analog chemistry program is initiated to synthesize numerous compounds in an attempt to find the best compound—usually as measured by efficacy. This procedure permits discovery of compounds with broad-spectrum activity, for example, preemergence herbicidal activity on broadleaf weeds, fungicidal activity against phycomycetes, or insecticidal activity against lepidopterous larvae. The industry has the expertise, personnel, and facilities to synthesize and screen thousands of chemical compounds and quickly assess their potential usefulness. Past successes provide adequate motivation and incentive to continue this approach.

Because of their host specificity, microbial pesticides, especially microbial herbicides, do not fit this traditional approach to discovery. To date, potentially useful microbial pesticides have been discovered by scientists who have studied a particular pest at an on-site location in search for an indigenous or exotic pathogen of that pest. Industry currently does not have the personnel, resources, or incentive to discover a pest control agent in this manner. Consequently, during this transition of reducing emphasis on chemicals and increasing emphasis on biological control agents, nonindustrial institutions such as universities, government agencies, and research foundations must play a significant role in the discovery of biological control agents (10,13). Several factors support this view: (1) most of the biological control agents discovered to date have been discovered by nonindustrial scientists; (2) there is a wider geographic distribution of scientists supported by the universities and federal agencies who have on-site involvement with several biological pest control systems than can be supported by industry; (3) scientists at a local level have a much greater opportunity to discover specific biological control agents because of their intimate knowledge of particular pest problems, uniqueness of pest biology, parasites or antagonists of the pest, and the environmental parameters that affect performance of the biological control agents; and (4) current economic considerations, particularly market size, host specificity, and other restraints, do not permit expenditures by industry on both discovery and commercialization of biological control agents. Thus significant inputs of public and research foundation funds will be necessary to discover and evaluate biological control agents. Industry may play a future role in the discovery of biological control agents; however, the role most likely will be directed toward selection of new strains or biological alteration of known microorganisms rather than finding previously undiscovered indigenous or exotic biological control agents.

IMPORTANCE OF PATENTS

Even with the input of public monies for discovery and initial evaluation of microbial pesticides, significant additional capital must be invested by industry to develop and market the discovery. Historically, a clear patent position has been necessary for an industry to invest substantial capital for research, development, and marketing of a product. A patent, therefore, may often be the key in an industry's involvement with development and marketing of a specific microbial biological control agent. The philosophy within the industry on the degree of patent protection required varies widely according to the particular project and the risk involved.

Patenting of the microorganism has been difficult in the past because the U.S. Patent and Trademark Office has maintained that microorganisms are not patentable. However, the U.S. Supreme Court recently ruled that novel forms of living organisms can be patented (15). As a result, a microbial pesticide may be protected by different types of patent: (1) a patent on the microorganism; (2) a patent on a composition of matter produced by a microbial culture; or (3) a process patent that may claim a method for production of the organism or its use, as in controlling weeds or other pests. Thus the strongest patent protection for microbial pesticides may now be possible. The subject of patenting microbial discoveries has been discussed in detail by Marcus (16) and Saliwanchik (17).

The importance of protecting discoveries with patents may not be recognized by nonindustrial biological scientists. Patents are commonly obtained by chemical and engineering scientists at universities and federal government agencies, but biological scientists often lack familiarity with patents.

Academic scientists have generally accepted their obligation to publish research results. Fortunately, disclosure of results or ideas through publication in scientific journals and obtaining a patent are not mutually exclusive; patenting simply requires proper timing. In the U.S. patent system, a patent application must be filed within 1 year after first public disclosure of the invention. If the application is not filed within this time, the information is considered to reside in the public domain, and the entity, use, or process is no longer patentable. In some countries, public disclosure is not permitted before the patent application is filed. Generally, initiation of patent proceedings at the time of submission of a manuscript to a scientific journal provides adequate time to protect patentability. Oral presentation and abstracts of papers at scientific meetings may constitute public disclosure. Therefore, it is prudent to refrain from presentations until sufficient data are accumulated to warrant filing a patent application.

Because patent policies vary among institutions, scientists should become familiar with institutional policies and be aware of procedures for filing, ownership, and division of royalties. In some cases, an industry may be willing to apply for a patent on behalf of an institution, provided that the industry is granted the right of first option to an exclusive license. Also, if industrial grant funds were used to support research leading to the patent, joint ownership of the patent may be possible. Since the source of funds used to support research affects ownership of patents, the use of public funds for private ownership of patents is a sensitive and controversial issue. When state or federal funds are used for discoveries in the United States, ownership of the patent resides with the institution or agency where the research was done. However, it is generally required that patents on discoveries made with funds from regional research projects, the USDA Competitive Research Grants Office, and USDA cooperative agreements be assigned to the U.S. Secretary of Agriculture for public use. In the latter case the value of a patent may be negated because exclusive licensing is precluded or at least discouraged. However, the federal government is relinquishing more and more of these patent rights, and university-affiliated scientists are now allowed to obtain patents on behalf of their university. Federal agencies such as the U.S. Department of Energy, the National Science Foundation, the National Institutes of Health, and the National Aeronautics and Space Administration can now give private companies exclusive licenses that permit royalty payments to cooperating university scientists. It is prudent, therefore, for the nonindustry scientist to consider the patentability of potential microbial control agents and not rush to publish immediately following discovery.

COLLABORATION BETWEEN INDUSTRIAL AND NONINDUSTRIAL SECTORS

As discussed previously, scientists in universities, government agencies, and research foundations will continue to conduct much of the early research directed toward understanding the biological control potential of microorganisms. If the potential biological control agents are to be useful, industry, university, and government scientists must collaborate. As Prager and Omenn (18) reported, "the generation of new knowledge and the translation of that knowledge into commercial products and services must be linked. Such linkage depends on close interaction between those who perform basic research (universities) and those for whom the results of basic research are the raw materials for product development and commercialization (industry)." Thus this interface between industrial and nonindustrial scientists is an essential key to accelerating progress in this type of pest control.

If the market potential for repeated applications of biological control agents to substantial hectarages of host populations provides an adequate profit incentive, industry will focus its sophisticated fermentation expertise and facilities on this type of research. But only after a rather extensive initial evaluation of the potential biological control agent will the industry commit fermentation facilities and collaborate with nonindustrial scientists to provide quantities of the organism sufficient for field testing.

It is important for industrial and nonindustrial organizations to recognize early and accept that only a few microorganisms with pesticidal activity may be suitable for commercialization. In comparison, only about 8 of some 8000 known antibiotics have been marketed. Moreover, for microbial pesticides to be accepted by the agricultural community, the pesticide product must be of high quality, efficacious, safe, and economical. When a microorganism is found that is likely to meet these qualifications, the collaboration between industry and nonindustry can be formed and the project advanced to the lead development phase.

LEAD DEVELOPMENT

Once a discovery has been made, many factors must be studied to determine whether that discovery can be commercialized. This period of study, called the *lead development phase*, is devoted to a better understanding of technological and economic factors leading to a decision on whether a particular microbial pesticide should be commercialized. During the early years of biological control technology, lead development will be based on a strong collaborative effort between the industry and nonindustrial scientists—probably much more so than has been the case with chemical pesticides.

TECHNOLOGICAL CONSIDERATIONS

Efficacy. The first question asked by most people involved with development of pest control strategies is, "will the invention control the target pest(s)?" For a biological control agent to be accepted by the user, it must be efficacious. Efficacy can be influenced by many factors, such as formulation, timing of application, placement of inoculum, and environmental conditions. Consequently, in the case of microbial herbicides, during lead development the actual level of efficacy demonstrated by the microorganism is not as critical as the potential level of efficacy.

There are very few economically important pests for which no control measures are available. As a result, from a commercial standpoint it is important to measure efficacy in terms of absolute and comparative values and according to how well the biological control agent compares with other available control measures.

Safety and Specificity. Safety and specificity questions cover a wide gamut and include mammalian toxicity, user and handler safety, environmental safety, and safety to nontarget organisms. It is impossible to investigate all parameters of safety and specificity during lead development. However, a primary concern is for the safety of the people working with the discovery. Consequently, the level of acute hazard of the microorganism, including acute oral LD_{50}, eye and/or skin irritation, inhalation, toxin potential, and tissue and blood responses would be established early in development.

Specificity is a double-edged sword. Specificity is highly desirable and is a principal advantage of biological control agents (5). However, specificity can also be an undesirable characteristic from a commercial standpoint if the microorganism is specific to only one pest and consequently has a small market potential. For example, the major reason for the delay in development of CGA has been the reluctance by industry to invest significant amounts of money on a biological control agent that controls only one weed in two crops in one state. Development would have progressed more rapidly if the fungus had activity against several *Aeschynomene* species or if northern jointvetch were a more widespread problem. Therefore, because of the importance of specificity to market potential, questions relative to pest spectrum, crop usage, and the effects on nontarget organisms are also examined in lead development. The development and registration costs for a product with small market potential and for a product with much larger market potential are essentially equal (13).

Other safety and specificity parameters, especially the long-term chronic toxicity and environmental safety studies, which require a commitment of at least 3 years and $1 million, will not be initiated until after a decision has been made to develop a marketable product.

Genetic Stability.   Whereas chemical pesticides are not subject to spontaneous generation, microbial control agents may be capable of reproduction in the environment. Consequently, it is critical that the biological control agent maintain the desired genetic constitution during mass production and/or after dispersal into the environment and not mutate into a form of reduced virulence or expanded host range.

Potential for Mass Production.   For a biological control agent to be commercially successful, it must be mass-produced at a cost that is economical to the user. As a practical matter, current industrial fermentation favors production of microorganisms in submerged liquid culture (see Chapter 9). Although mass production of obligate parasites or organisms that require solid substrates is biologically feasible, substantial and expensive technological developments will be required before solid-state (surface) fermentation supplants the submerged fermentation facilities and technology now available in the pharmaceutical, yeast, and brewing industries. Solid culture of bacteria and fungi may be suitable to assess initial activity and specificity, but production of large quantities needed for field tests usually cannot be produced by these techniques. Thus for commercial development of a biological pesticide, growth and sporulation of the organism in submerged liquid culture may be an essential requirement.

This generalization is not valid for insect viruses and protozoa as demonstrated by the commercial production of polyhedral inclusion bodies of insect NPV and the protozoan *Nosema locustae*. Great strides are being made toward production of insect viruses by cell culture techniques. For commercialization, it would be desirable if this technique could be developed to a point where viruses, protozoa, and other obligate parasites could be mass-produced in cell culture by techniques similar to submerged fermentation.

Formulation.   Formulation is the blending of an active moiety, in this case a microorganism, with diluents and surface-active agents (surfactants). The purpose of formulation is to alter physical characteristics to a more desirable form such as to dilute to a common potency, to enhance stability and/or biological activity, or to improve mixing and sprayability (19). Numerous additives and diluents must be tested singly and in various combinations to obtain a good formulation. Although there is considerable knowledge regarding formulation of chemical pesticides, formulation of living entities is a relatively new science. Usually in the pharmaceutical industry the microorganisms produced from fermentation are killed, and then the drugs are extracted and formulated. Conversely, for most biological control agents, the life and virulence of the organisms must be maintained.

The Upjohn Company has encountered two difficulties in the formulation of CGA, both involving the maintenance of viability: (1) many adjuvants were found to be toxic to the spores; and (2) vigorous physical manipulation killed most of the spores. Through time, perseverance, and innovation these obstacles to formulation have been overcome.

Stability and Shelf Life.   Any product, whether biological or chemical, must maintain its physical, chemical, and/or biological integrity from inception through final use. This period can vary from a few months to several years. Consequently, the microorganism must be produced in a stable form, such as spores, sclerotia, or resting cells. A manufacturer has control over handling and storage conditions when the product is being produced and packaged. However, the manufacturer cannot control the product after it has been released to distributors, dealers, and users. Therefore, it is important that the biological control agent maintain viability over a reasonably wide range of handling and storage conditions.

ECONOMIC CONSIDERATIONS

Industry is not reluctant to spend a substantial amount of funds on a venture that may be considered risky. However, industry can survive and grow only if a profit can be made. It is essential, therefore, that preliminary economic analyses be conducted before a significant amount of funds is invested.

Market Potential.   After determining the spectrum of pests controlled by a biological control agent and its crop selectivity, the potential usefulness of this agent can be assessed. Numerous publications are available that list crop hectarages in the world. The size of a market can be ascertained from these publications, although a major difficulty is the estimation of the share of that market that can be obtained. This is especially a problem for microbial pest control agents that encompass a new concept.

As for CGA, it was known early that the fungus was useful in controlling northern jointvetch in rice (8). Later it was learned that CGA would also control the weed in irrigated soybean. Data were readily available that indicated rice and soybean hectarages in eastern Arkansas, the only area where northern jointvetch is a problem. From these data the market size was established. Then came the task of ascertaining (1) the hectarage on which northern jointvetch was a problem, especially after use of standard preemergence herbicides, (2) the hectarage on which postemergence herbicides were used, and the target weeds for these herbicides, (3) the hectarage on which northern jointvetch was the only postemergence weed problem, (4) the postemergence herbicides that were being used to control northern jointvetch and why they were being used, and (5) the hectarage on which CGA would be used. Projecting the future is risky. Nevertheless, a market potential for CGA was established. Only with time will the accuracy of the estimated market potential be determined.

Research and Development Costs.   Industries that are experienced in chemical pesticide development are aware of research and development (R&D) costs. Discussions on the time frame and costs involved in pesticide R&D are available (13,20). Until 1978 the EPA was requiring the same types of data for registration of biological pesticides as were required for chemical pesticides. This made it easier for the industry to estimate reliably the R&D costs for biological pesticides. Since 1978 the EPA has been writing new guidelines, and consequently the pesticide industry has been somewhat uncertain as to what data will be required and how much these will cost to generate for both chemicals and biologicals. In August 1980 the EPA proposed newer guidelines (summarized in Chapter 11) for registration of biorational pesticides (1). From these guidelines, an estimate of development time and costs can be projected.

Capital Investment. One major attraction to the fermentation industry for commercial development of microbial biological control agents is that production of these agents is compatible with the technology and equipment existing in the industry. However, additional equipment and production facilities may be needed depending on market potential, quantity of product needed, and biological properties of the organism (see Chapter 9). It is important to speculate early in the lead development phase a procedure for mass production and the estimated need and cost of capital investment.

Product Cost. As with the other economic considerations, the cost of the pesticide product is highly speculative during lead development. Nevertheless, on the basis of past experience, knowledge of fermentation technology, and available data, a projected product cost must be established. The affordable price of a pesticide will be determined not by the cost of goods, but by the value of the crop protected (13) or of the pest controlled. Thus if the product cost is too great, there will not be enough profit for the industry to recoup its investment.

Royalty. As discussed previously, universities and government agencies appear best suited to discover and initially evaluate potential biological control agents. In return for the right to use that technology, especially an exclusive right, the originating institution or agency may receive a royalty. The amount of royalty depends on the ultimate value of the technology plus the amount of data generated by the licensing agency versus the amount of data necessary to commercialize the invention.

Return-on-Investment Analysis. To estimate the potential profit to be derived from a project, a return-on-investment (ROI) analysis is conducted; ROI is a numerical estimate of the dollar value of a project. This estimate allows the project to be compared on a common basis with other corporate projects competing for a finite amount of funds and personnel. Many factors are included in ROI analysis (Fig. 1), but in simple terms the ROI analysis is a measure of the estimated cash inflow versus the estimated cash outflow over a period of time. The ROI value is calculated as a percentage return on the cumulative cash flow. The effects of various factors on the cumulative cash flow have been discussed by Riggleman (21).

The ROI analysis does not eliminate the use of subjective evaluation of a project such as the probability of technological and marketing successes,

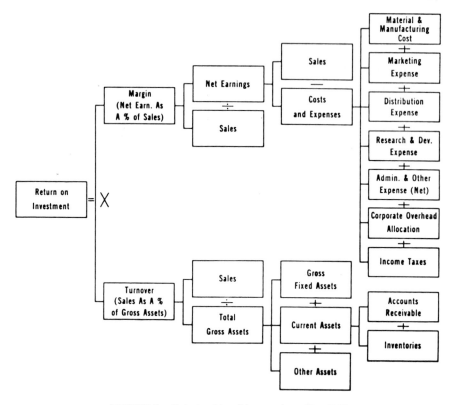

FIGURE 1.   Relationship of factors that affect ROI.

structuring the knowledge and technological bases for future projects, better use of existing equipment and personnel, and societal benefits to be derived from the project. For example, the information obtained from CGA is considered as a foundation that can be used to develop other commercial biological control agents.

Finally, lead development is not a short-term effort; it continues over a period of one to several years. As technological and economic data are generated, the project is frequently updated and reanalyzed. At some undefinable point, a decision is made to either discontinue the project or to continue it and advance to the product development phase.

PRODUCT DEVELOPMENT

Product development is in many ways a continuation and expansion of the lead development phase. Data continue to be gathered that relate to the technological and economic considerations discussed previously. However, there are some significant differences. The product development phase requires commitment by the industry to invest substantial money, time, and personnel. At this juncture a carefully considered decision must be made because, from here on the industry will assume total risk for that product. Equally important, there is a significant change in attitude—a change from one of purely scientific curiosity to one of commitment to develop and market a product. Any concept, project, or product must have proponents from all sectors involved to be successful. This change in attitude results in a redirection of goals. No longer are data gathered only for scientific purposes; rather, data are now generated to develop a use pattern for a high-quality product that is efficacious, safe, and economical.

DEVELOPMENT OF USE PATTERN

Since a biological control agent developed in one geographic area may not perform satisfactorily in another area, large-scale field trials are conducted at many locations to evaluate the agent under varying environmental conditions. Test results from several areas provide data on the effect of temperature, humidity, rainfall, soil type, equipment, and local crop production practices on the performance of the biological control agent. Rate, timing, and method of application studies are included not only to confirm results of earlier small-scale efficacy tests, but also to establish the optimum conditions for maximum effectiveness. In the case of microbial herbicides, as discussed in Chapter 8, many factors related to the development of epiphytotics must be identified and quantified.

Large-scale field tests also provide information on how the biological control agent should be integrated with existing pest management programs (see Chapter 12). For example, the use of CGA to control northern jointvetch must be integrated with chemical herbicides that are used for controlling other weeds in rice and soybean. More importantly, CGA must also be integrated with the fungicide benomyl, which is commonly used in rice culture and is toxic to CGA spores.

Other factors studied in the large-scale field tests include evaluation of the product with respect to its safety to nontarget plant and animal species, efficacy compared to competitive herbicides, effects on crop yield and quality, use limitations such as when and where the product should not be used, and effect of the product on the environment.

Numerous field trials are a requirement of any reputable industry. These field trials are necessary to determine the risk the company and society will be assuming when a product is marketed. The field trials are also used to teach the sales personnel, distributors, dealers, and users how to use the product to achieve maximum benefits.

SAFETY PARAMETERS

In its guidelines (1) on data required for registration, the EPA has proposed a tier testing approach to ascertain the safety of biorational pesticides (see Chapter 11). If adopted, this scheme would eliminate the need for extensive and expensive studies on residue chemistry, toxicology, nontarget organism hazard, and environmental fate for those microbial pesticides that are determined to be safe on the basis of Tier I testing. The EPA anticipates very few tests beyond Tier I because most questions concerning hazards of microbial pesticides are expected to be answered by the first tier of tests (1).

All safety tests must utilize the pesticide product in the form in which it will be marketed. Consequently, most of the mandated tests must be conducted after the manufacturing procedure and product formulation have been finalized.

FINAL PRODUCT

Biological control agents must be efficacious, safe, and of high quality to be accepted by the users. Therefore, quality assurance assays and specifications must be developed. For chemical pesticides, quality assurance parameters generally can be measured by routine chemical analyses, whereas, for most of the microbial pesticides, specific bioassays must be developed. These bioassays must be accurate, reliable, rapid, and economical. Moreover, the assays must measure the potency and purity so that the microorganism of choice can be distinguished from contaminants. For example, an assay had to be developed for CGA that would indicate potency as measured by number of viable spores per gram. In addition, an assay had to be developed to distinguish CGA from other fungi including *Colletotrichum gloeosporioides* f. sp. *jussiaeae*, another microbial herbicide under development by The Upjohn Company and that is specific to winged waterprimrose (*Jussiaea decurrens*).

Quality assurance also involves delivery of a product to the user with the levels of potency and purity that are guaranteed on the label. To fulfill this obligation, data must be obtained and specifications written relative to product stability, shelf life, packaging, transportation, and storage. Since the product in this case contains a living entity, the handling, transportation, and storage conditions probably will be more critically defined than for most chemical pesticides. Special treatment of the product such as transportation and storage in refrigerated containers may be necessary. Also, short-term expiration dating of the product may be required. These requirements may cause manufacturing and marketing problems concerning scheduling of production, distribution, and possibly product recall and destruction after the expiration date.

As it becomes clearer that a final product may be a reality, procedures for mass production must be finalized as discussed in Chapter 9 and by Kenney and Couch (10). Frequently, these procedures may entail modification of the production plant or construction of a new facility. In such cases, the modification or the construction will be initiated during the product development phase to ensure expedient production and marketing of the product.

Industry has the responsibility to submit all data required for registration to the EPA and to support and defend these data if necessary. Also, the product must be registered in all states in which it will be sold. State requirements vary widely. Some states accept the EPA registration, but certain other states require an additional, in-depth review that is independent of the EPA review. California and New York are examples of the latter. After federal and state registration requirements have been met and a label has been approved, the product can be sold (see Chapter 11, section on interstate movement of plant pathogens).

CONCLUSIONS

A transitional period of several years will probably pass before microbial pest control agents are fully accepted for agricultural use. During this transitional period, universities, government agencies, and private foundations will shoulder a significant responsibility for the discovery, identification, and early lead development of microorganisms that possess pesticidal properties. When a microorganism is discovered that has the potential to become an efficacious, safe, economical, and high-quality product, a collaboration between nonindustrial and industrial organizations must be formed to advance the discovery to a useful product. Industry will attempt to commercialize those microbial pest control agents that show promise of being profitable.

Sincere appreciation is extended to W. Klomparens and H. R. Keyser of The Upjohn Company, Kalamazoo, MI, and G. E. Templeton, University of Arkansas, Fayetteville, AR, for their valuable assistance in preparation of this manuscript.

REFERENCES

1. Anonymous. 1980. *Proposed Guidelines for Registering Pesticides in the United States.* Subpart M. *Preamble and Guidelines.* August 4, 1980. The Environmental Protection

Agency, Hazard Evaluation Division, Office of Pesticide Programs, Washington, DC, 406 pp.

2. Anonymous. 1979. *Proposed Guidelines for Registering Pesticides in the United States. Subpart G. Preamble and Guidelines.* June 22, 1979. The Environmental Protection Agency, Hazard Evaluation Division, Office of Pesticide Programs, Washington, DC, 676 pp.

3. Anonymous. 1979. *Pest Management Strategies in Crop Production.* Vol. 1. Report OTA-F-98. Congress of the United States, Office of Technology Assessment, Washington, DC, 132 pp.

4. R. W. Hall, L. E. Ehler, and B. Bisabri-Ershadi. 1980. Rate of success in classical biological control of arthropods. *Bull. Entomol. Soc. Am.* **26**, 111–114.

5. E. W. Lawless, R. von Rumker, G. L. Kelso, K. A. Lawrence, J. D. Maloney, and E. R. Thompson. 1977. *A Technology Assessment of Biological Substitutes for Chemical Pesticides.* Report PB-284-990, National Science Foundation, U.S. Department of Commerce, National Technical Information Service, Washington, DC, 407 pp.

6. Anonymous. 1975–1976. *Pest Control: An Assessment of Present and Alternative Technologies.* Vols. 1–5. National Academy of Sciences, Washington, DC.

7. Anonymous. 1977. *New, Innovative Pesticides: An Evaluation of Incentives and Disincentives for Commercial Development by Industry.* Stanford Research Institute, Menlo Park, CA, 266 pp.

8. J. T. Daniel, G. E. Templeton, and R. J. Smith, Jr. 1974. Control of *Aeschynomene* sp. with *Colletotrichum gloeosporioides* (Penz.) Sacc. f. sp. *aeschynomene.* U.S. Patent No. 3,849,104.

9. D. S. Kenney, K. E. Conway, and W. H. Ridings. 1979. Mycoherbicides—potential for commercialization. *Dev. Ind. Microbiol.* **20**, 123–130.

10. D. S. Kenney and T. L. Couch. 1981. Mass production of biological agents for plant disease, weed and insect control. Pp. 143–150 in: G. C. Papavizas, ed., *Proceedings of the Beltsville Symposia in Agricultural Research. 5. Biological Control in Crop Production.* Allanheld, Osmun and Company, Totowa, NJ.

11. T. L. Couch. 1979. Standardization of entomogenous fungi. Pp. 138–146 in: C. M. Ignoffo, ed., *Proceedings of the First Joint US/USSR Conference on the Production, Selection, and Standardization of Entomopathogenic Fungi. Project 5. Microbial Control of Insect Pests.* The US/USSR Joint Working Group on the Production of Substances by Microbiological Means. JURMULA, Riga, Latvia, USSR.

12. M. H. Rogoff. 1973. Industrialization. In: L. A. Bulla, Jr., ed., Regulation of insect populations by microorganisms. *Ann. N. Y. Acad. Sci.* **217**, 200–210.

13. C. Djerassi, C. Shih-Coleman, and J. Diekman. 1974. Insect control of the future: Operational and policy aspects. *Science* **186**, 596–607.

14. C. M. Ignoffo and T. L. Couch. 1980. *Baculovirus heliothis*: The nucleopolyhedrosis virus of *Heliothis* species. Pp. 327–360 in: H. D. Burges, ed., *Microbial Control of Insects, Mites, and Plant Diseases.* Vol. 2. Academic Press, London.

15. S. Diamond. 1980. Commissioner of Patents and Trademarks v. Chakrabarty, A. M. Supreme Court of the United States. *U.S. Quarterly* **206**, 193–202.

16. I. Marcus. 1979. Fermentation process and products: Problems in patenting. Pp. 497–530 in: H. J. Peppler and D. Perlman, eds., *Microbial Technology*, 2nd ed. Vol 11. Academic Press, New York.

17. R. Saliwanchik. 1976. Bugs and patents. *Dev. Ind. Microbiol.* **17**, 135–138.

18. D. J. Prager and G. S. Omenn. 1980. Research, innovation, and university-industry linkages. *Science* **207**, 379–384.

19. E. A. Snow. 1980. The Upjohn Company, Kalamazoo, MI, personal communication.

20. C. H. Gilbert. 1978. Industry's view on benefits and risks of the pesticide business. *Proc. South. Weed Sci. Soc.* **31**, 14–27.

21. J. D. Riggleman. 1979. The cost of regulation. *Agrichem. Age* **23** (4), 6–8.

11

REGULATION OF MICROBIAL WEED CONTROL AGENTS

R. CHARUDATTAN

Plant Pathology Department, University of Florida,
Gainesville, Florida

Some proposed and current regulations concerning the importation, registration, and use of plant pathogens as weed control agents are discussed in this chapter. Some of the proposed regulations are subject to revisions; therefore, this chapter presents a summary of the general scope and purpose of the regulations. Specific details can be found in the references cited or obtained from the agencies mentioned. Although biological control of weeds with plant pathogens is a field that is still in its infancy, regulations are deemed necessary both to ensure safety to the public and to encourage the use of plant pathogens in weed management systems. It is generally accepted that biological control agents are fundamentally different in their properties and mode of action from chemical herbicides and owe their effectiveness to phenomena other than innate toxicity. Consequently, they may pose a lower risk of adverse effects on the environment than do conventional pesticides. However, before introducing an exotic weed pathogen into a new environment or augmenting an indigenous pathogen in its native habitat, it is prudent to establish the specificity, efficacy, and lability of the microbial weed control agent. In this respect, regulatory guidelines can suggest the most expedient and proper testing scheme to follow for evaluation of a microbial weed control agent. Regulations also can serve as a legal base for protection against

Florida Agricultural Experiment Stations Journal Series No. 2952.

misuse and false claims involving microbial agents. However, excessive regulation will stifle progress in this field through delays and high cost of compliance. Moreover, regulations and guidelines should be flexible to allow for biological differences among plant pathogens and the pathogen–host systems involved in biocontrol.

Much of the work on weed control with plant pathogens is being done in the United States (1), where the pertinent regulations are more well defined than elsewhere. Therefore, this chapter is devoted mainly to the American examples, with a brief discussion of the Australian regulations.

Pathogens may be used as either classical biological controls or as microbial herbicides (see Chapter 3 for definitions), and pathogens may be exotic or native to the country in which they are used. Historically, classical biological control agents have been exotic organisms, whereas native or indigenous pathogens have been used as microbial herbicides. Obligate parasites such as the rust fungi have been used more commonly as classical biological controls, whereas facultative parasites, whether native or exotic, are best suited for use in the microbial herbicide strategy. In the United States, the types of regulation imposed on the use of pathogens as biological weed control agents depend on the geographic origin of the pathogens. The importation and release of exotic pathogens is authorized and regulated by federal and state agencies (see section on importation of exotic pathogens). Both exotic and native pathogens, used either in the classical tactic or as microbial herbicides, are regulated and registered by the federal government (see section on microbial herbicides). However, provisions exist for exempting certain biological control agents from this general rule (see section on biorational pesticides).

IMPORTATION OF EXOTIC PLANT PATHOGENS
FOR BIOLOGICAL CONTROL OF WEEDS

Surveys for plant pathogens in the native range of imported weeds have been useful in finding pathogens suitable for biological control. *Puccinia chondrillina* and *Phragmidium violaceum* are good examples of introduced pathogens that have controlled two imported weeds, respectively, in Australia and Chile (2,3). Foreign explorations for insect biological controls of imported weeds have also been very successful in many instances (4,5). For certain introduced weeds, native biological controls may not be available or suitable; only foreign organisms may help in this situation. Because the potential for success in biological weed control with exotic plant pathogens is good, the value of foreign explorations must not be underrated. Regulations that encourage foreign exploration and importation of promising pathogens are essential to the success of any biological weed control program.

IMPORTATION INTO THE UNITED STATES

In the United States, importation and release of exotic plant pathogens for biological control are regulated. The Plant Quarantine Act of 1912 (6) and the Federal Plant Pest Act of 1957 (7) prohibit the importation of foreign plant pathogens without authorization from the U.S. Department of Agriculture (USDA). The Plant Protection and Quarantine (PPQ) section of the USDA's Animal and Plant Health Inspection Service (APHIS) issues permits for such imports. Generally the Department of Agriculture of the state into which the pathogen is to be imported is contacted with an import application (PPQ form 526, available from USDA, APHIS, PPQ, Plant Importation and Technical Support, Hyattsville, MD 20782). In the case of pathogens of weeds, the PPQ solicits the help of a working group (8), established jointly by the USDA and the U.S. Department of the Interior, to review the application. The group, called the Working Group on Biological Control of Weeds (WGBCW), is comprised of scientists from federal agencies concerned about the effects of introducing exotic organisms into North America. The WGBCW also reviews proposals on request by the researcher and provides recommendations to the PPQ and the researcher on the testing and release of biological agents. The researcher may request the WGBCW to review proposals before initiating foreign surveys on a target weed to be advised on potential conflicts of interest (whether the weed should be controlled and whether foreign surveys should be undertaken). In this case, the WGBCW may suggest a list of hosts to be screened in specificity testing.

Whenever practical, a potential biological control agent is screened and evaluated in its native habitat before importation into the United States. For example, the introduction of *P. chondrillina* into the United States (9) was facilitated because it had been thoroughly evaluated in its native range in Europe (10,11) and demonstrated to be a successful biocontrol agent in Australia (2). The USDA maintains overseas biological control laboratories, primarily for entomological studies, where plant pathogens may be screened. Alternatively, collaboration with scientists of the Commonwealth Institute of Biological Control (12), the International Organization for Biological Control (12), and other agencies involved in biological control (12) can assist in foreign surveys and screening.

Exotic pathogens may be imported into an approved quarantine laboratory in the United States for evaluations of host range, safety, and efficacy. This approach was used in Florida (13) for screening plant pathogens of waterhyacinth *(Eichhornia crassipes)* and hydrilla *(Hydrilla verticillata)*. The approval of construction and operation of quarantine facilities and the determination of the adequacy of the facilities and the technical competence of investigator(s) are the responsibility of the PPQ and the State Departments of Agriculture (SDA). Presently two facilities are involved with research on exotic plant pathogens for biocontrol of weeds (14). The USDA operates one facility in Frederick, Maryland, and the University of Florida operates the other in Gainesville, Florida.

Certain minimum precautions are warranted in the handling of exotic pathogens during testing and release. Species identification of the pathogen and preliminary host range information are desirable prior to introduction into quarantine facilities. Any shipment of pathogens to the United States must be in containers meeting PPQ safety standards and shipped to approved quarantine facilities under the appropriate PPQ permits. Voucher specimens of the weed and the pathogen should be deposited, respectively, with the U.S. National Arboretum Herbarium, Washington, DC 20002 and the Plant Disease Laboratory, Frederick, MD 21701. The pathogen should not be distributed to other researchers or laboratories without approval from the PPQ. If a pathogen is found to be a potential pest of valued plants, cultures of the organism should be destroyed.

Once an exotic pathogen has been found to be a desirable biocontrol agent, a proposal is made by the researcher to the WGBCW for review and recommendation for its release from quarantine. In reviewing the proposal, the WGBCW will be concerned primarily with the safety of the proposed introduction. The host range data provided by the researcher forms much of the basis for approval or disapproval of the pathogen. Efficacy of the pathogen and the economic need for biocontrol (importance of the weed problem) are secondary, but they also are important considerations. The WGBCW will concur with the Canadian and Mexican regulatory agencies before recommending release of an exotic plant pathogen in the United States. If the WGBCW recommends that the pathogen be released for field use or testing, the PPQ and the SDA may authorize such release. An exotic pathogen may not be initially cleared for mass release but might be approved for limited field tests. Further releases will often depend on success of these initial tests.

This overall process of screening and evaluating potential exotic biocontrol agents for release in the United States has served well. The WGBCW has played a valuable role in the regulation of exotic biocontrol agents, including plant pathogens.

IMPORTATION INTO AUSTRALIA

Importation and release of exotic plant pathogens into Australia are authorized by the Plant Quarantine Section of the Commonwealth Department of Health. A Culture Committee and, if necessary, a select panel of experts are consulted before a decision is made. Following the importation of *P. chondrillina*, which lead to the spectacularly successful control of skeletonweed (2), a proposal to import a *Chondrilla* form of the powdery mildew *Erysiphe cichoracearum* was refused (15). Subsequently, in 1978 the Australians established a working group to suggest guidelines for importation of plant pathogenic fungi (15). The group consists of plant pathologists, representatives from the Department of Health, the Commonwealth Scientific and Industrial Research Organization (CSIRO), and state departments concerned with biological control of weeds.

The Australian and American guidelines on importation and release of plant pathogens are similar, except that the Australian guidelines are restricted to fungi only. Both specify the types of information needed in support of proposals for importation and release. For example, proposals should discuss the taxonomy of the weed and the plant's importance, biology, habitat, and the potential for its biological control. Likewise, the pathogen's taxonomy, detrimental (potential for mycotoxin production) and beneficial aspects (potential for biological control), and other known hosts, if any, should be indicated. Information on the pathogen's biology, reproductive stages, worldwide distribution, native range, and center of origin, and the epidemiology of the disease it causes is also required.

The Australian guidelines (15) propose a host range testing scheme for determining the specificity of the pathogen. The CSIRO maintains overseas biocontrol laboratories where such testing is done, because, according to these guidelines, all testing must be done outside Australia, by personnel and under conditions approved by the Australian Department of Health. The environmental conditions for host testing, types of plant to be tested, and agencies (Agriculture, Forestry, CSIRO, etc.) to be consulted are also specified in the guidelines. The fungal isolate selected for release (should be the same isolate used in the specificity and efficacy evaluations) should be identified by its varietal or strain designation and assigned a number. The fungus need not be a single-spored strain. A mixture of strains is preferred, not only for increasing the prospects of long-term control of the weed (through effectiveness against several forms of the weed and by recombination of the virulent strains), but also for indicating the potential for the pathogen's variability (see Chapter 7). A culture or specimen of the fungus must be deposited at the Commonwealth Mycological Institute, Kew, England, and another at the Department of Agriculture, Biology Department, Rydalmere, New South Wales, Australia. Fungicides capable of controlling the fungus must be selected.

Information on the regulation of exotic plant pathogens for weed control in other countries has been difficult to obtain. It is assumed that provisions exist in the plant quarantine laws of every country for regulation of the use of plant pathogens as biological control agents. If not, the examples cited here may be useful in establishing regulatory guidelines. International organizations involved with biological control (12) and crop protection (16) have the opportunity to play a leadership role in developing these guidelines.

USE OF PLANT PATHOGENS AS MICROBIAL HERBICIDES

In the United States, microbial herbicides and other microbial pesticides are regulated and registered by the federal government. The Federal Insecticide, Fungicide, and Rodenticide Act (FIFRA) as amended (17) dictates the regulation of the pesticidal use of microorganisms and the registration of microbial pesticides. The FIFRA also authorizes an administrator of the federal government, specifically the administrator of the U.S. Environmental Protection Agency (EPA), to publish guidelines specifying information that is required to support the registration of a pesticide and to revise the registration guidelines as needed.

In 1980 the EPA proposed a set of guidelines for registration of biological pesticides, which, pending review, will become a law (18). The guidelines discuss registration procedures, data and labeling requirements for registration, standards for product efficacy, acceptable test methods for the development of required data, and the information required in test reports (18).

BIORATIONAL PESTICIDES

The EPA has included certain biological control agents under the name of "biorational pesticides" (18). The name construes a "biological basis" (as opposed to a chemical basis) for this group of pesticides. It does not, however, denote any "irrationality" in the use of other pesticidal agents; such as chemicals, which will continue to be a part of pest control strategies. As defined by the EPA, biorational pesticides are "often referred to as biological control agents, are chemical or living microscopic pesticidal agents, (and) are inherently different from conventional pesticides" (18). Furthermore, the chemicals referred to are certain naturally occurring biochemicals, and the microbial pest control agents include bacteria, rickettsias, mycoplasmas, fungi, protozoa, viruses, and members of the Class Schizophyceae. All other biological control agents are proposed to be exempt from the provisions of FIFRA (19). Also, provisions exist for exempting certain nonexempt biological control agents and subjecting others that are presently exempt to the FIFRA regulations (19). Questions concerning exemption and nonexemption will be resolved by the EPA (19).

In developing the guidelines, the EPA has taken a positive approach toward regulation of biological control agents. This is reflected in the statement, "EPA will ... take steps to substantiate by scientific data the expectation that many classes of biorational control agents pose lower risks than conventional pesticides. Although biorational pesticide registrants will not be relieved of the burden of proof of their safety, the (EPA) will take into account the fundamentally different modes of action of biorationals and the consequent lower risks of adverse effects from their use" (18).

SCOPE OF THE REQUIRED TESTING

A registrant must submit certain data on the nature or composition, safety, and efficacy of the biorational pesticide. The proposed EPA guidelines (18) have identified eight specific areas of testing and have established standards for testing and requirements for data submission. These areas include product analysis, residue chemistry, toxicology, nontarget organism hazard, environmental fate, product performance (efficacy), experimental use permit data, and label development.

In testing a biorational pesticide, a tier testing scheme is proposed, and the tests would be conducted in a manner that maximizes the pesticide's potential for hazard (maximum hazard testing). Tier testing is proposed to minimize the cost of generation of registration data. Instead of gathering extensive, across-the-board data for each biorational pesticide, the registrant would test in a "hierarchical system" (18). Table 1 illustrates the tier testing scheme for residue chemistry and toxicology and Table 2, for nontarget organisms and environmental expression. The Tier I tests will be required for all biorational pesticides except under residue chemistry. The decision to initiate Tier II tests will depend on the data from Tier I, Tier III on Tier II data, and so on. In most cases, questions concerning hazards of biorational pesticides should be answered by the Tier I tests (18). Tests up to four tiers are proposed.

Under the maximum hazard testing scheme, test target organisms will

TABLE 1. Summary of Proposed Tier Testing Scheme for Microbial Herbicides: Residue Chemistry and Toxicology

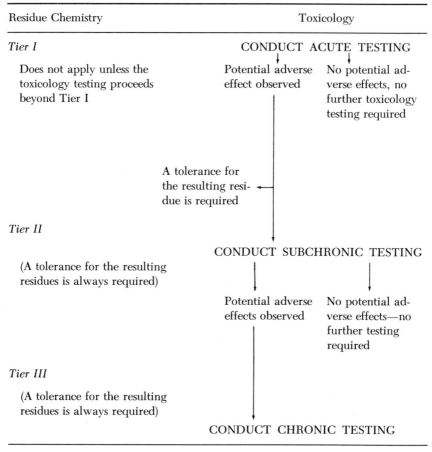

Residue Chemistry	Toxicology
Tier I	CONDUCT ACUTE TESTING
Does not apply unless the toxicology testing proceeds beyond Tier I	Potential adverse effect observed — No potential adverse effects, no further toxicology testing required
	A tolerance for the resulting residue is required
Tier II	CONDUCT SUBCHRONIC TESTING
(A tolerance for the resulting residues is always required)	Potential adverse effects observed — No potential adverse effects—no further testing required
Tier III	
(A tolerance for the resulting residues is always required)	CONDUCT CHRONIC TESTING

Source: Ref. 18.

be subjected at the Tier I level to extremely adverse exposures in terms of treatment dose or concentration, route of administration of the test material, and age of the test organism. It is believed that negative results with this kind of test at Tier I will provide a high degree of confidence in the safety of most biorational pesticides.

All eight areas referred to previously will not be subjected to the tier and maximum hazard testing. The residue chemistry, toxicology, nontarget organism hazard, and environmental fate would be tier tested. The

TABLE 2.   Summary of Proposed Tier Testing Scheme for Microbial Herbicides: Nontarget Organism and Environmental Expression

Nontarget Organisms   Environmental Expression

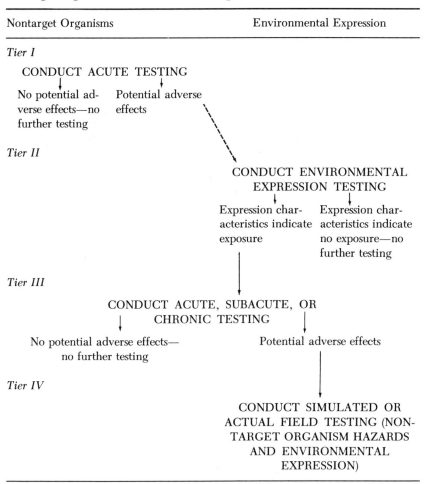

Tier I

CONDUCT ACUTE TESTING

No potential ad-   Potential adverse
verse effects—no   effects
further testing

Tier II

CONDUCT ENVIRONMENTAL
EXPRESSION TESTING

Expression char-   Expression characteristics indicate   acteristics indicate
exposure   no exposure—no
                       further testing

Tier III

CONDUCT ACUTE, SUBACUTE, OR
CHRONIC TESTING

No potential adverse effects—   Potential adverse effects
no further testing

Tier IV

CONDUCT SIMULATED OR
ACTUAL FIELD TESTING (NON-
TARGET ORGANISM HAZARDS
AND ENVIRONMENTAL
EXPRESSION)

Source: Ref. 18.

areas of toxicology and nontarget organism hazard would be tier tested and also subjected to maximum hazard testing.

Certain biorational pesticides may qualify for a limited level of testing (reduced data requirements) in certain of the eight test areas if the potential for human and other nontarget organism exposure and hazard from these pesticides is minimal. In other words, those biorational pesticides that are considered to be of lower risk to nontarget organisms qualify for reduced testing. However, if the reduced testing regimen indicates a hazard, further testing beyond Tier I may be required. Certain examples have been given (18). First, because of the extreme sensitivity of the target organisms, pheromones are applied at very low rates—rates at which hazards to humans and other nontarget organisms are unlikely. Therefore, it is proposed that biorational pesticides applied at rates of $\leq$ 49 g/ha per application may qualify for the reduced data requirements. Second, biorationals that are dispensed in tapes, traps, inert fibers, and other passive dispensers may be considered safe from direct exposure to humans and other nontarget organisms and thus may be exempt from certain testing requirements. Similar exceptions can be made for certain highly volatile biochemicals that would leave minimal or no residues at treated sites and thus pose little exposure hazard to nontargets.

The pesticides of concern here, the microbial herbicides, will be subject to Tier I testing. Toxicology data will be mandatory, but the need for data on residue chemistry will depend on the toxicology data. Data on nontarget organism hazard will be mandatory; the requirement of data on environmental expression (but not on environmental fate) will depend on the nontarget organism hazard data (Tables 1 and 2). In general, biorational pesticides intended for aquatic use may not qualify for reduced level of testing since water, unlike land, as a life-support medium is assumed to be far more vulnerable to damage.

PRODUCT ANALYSIS, PRODUCT PERFORMANCE, AND LABEL DEVELOPMENT

Microbial pest control agents, like conventional chemical pesticides, will require product analysis data, although the types of test will be different from those required for conventional chemical pesticides. As a minimum, microorganisms used as pesticides must be identified to the extent possible by taxonomic position, serotype, and strain. Composition of the pesticide formulations, additives, adjuvants, and inert ingredients must also be indicated. As for efficacy or product performance, two types of data will be required: the time required and the minimum effective dosage for a product to achieve the desired level of pest control or to meet its performance standard. Regarding the label requirement, the EPA recognizes the need for specific statements in the label proclaiming the lack of harmful effects from the biorational pesticide to beneficial organisms that may be vital to integrated pest management programs. It is conceivable that such statements may be allowed in the future (pending amendment to the existing law), but in the interim the label will have relatively fewer precautionary statements compared to chemical pesticides. Even "signal words" (18) expressing the lower degree of risk inherent in the biorational pesticides may be allowed, but no direct claim of safety to humans or nontargets can be made.

EXPERIMENTAL USE PERMIT

An experimental use permit (EUP) is granted for a potential pesticide undergoing experimental evaluation. The EUP system has been in use for all pesticides, and microbial herbicides are also evaluated under this system. Under the EUP, guidelines for evaluation and any limitations on testing of a potential pesticide are set forth. The data required in support of registration are specified. Before an application is made for an EUP, ordinarily the applicant will be expected to have information on how the microbial pesticide product is developed and data on product analysis. The basic toxicology and toxicity to nontarget organisms will be conducted under the EUP. Unless problems are indicated, no data on residues or environmental fate will be required. Also under the EUP, data on efficacy and effective dosage to achieve product performance standard will be gathered. A time frame for data submittal will be specified in the EUP.

REGULATION OF MICROBIAL HERBICIDES
IN OTHER COUNTRIES

Specific regulations on microbial herbicides in other countries are either nonexistent or in developmental stages since this approach to weed control has not been attempted in most countries. However, the experience gained from microbial insecticides such as *Bacillus thuringiensis* and the *Heliothis* nuclear polyhedrosis virus has resulted in regulations in countries where these insecticides are sold. It is understood that the Australian, British, Swedish, and West German regulations on microbial pesticides are similar to the American regulations but are generally more concise (20,21). It should be simple to adapt these existing pesticide regulations to microbial herbicides when the need arises.

INTERSTATE MOVEMENT OF PLANT PATHOGENS
USED FOR WEED CONTROL

Once registered by the EPA, microbial herbicides would be expected to be available for use anywhere in the United States. Current restrictions on the interstate movement of plant pathogens, which many states impose, would need to be modified, and it is unclear whether microbial herbicides will require state clearances for regional use. This aspect of interstate movement needs to be resolved soon. Plant pathogens of the classical biocontrol type, on the other hand, will most likely continue to require clearances from the USDA and SDA before deliberate movement across state lines.

IMPACT OF REGULATIONS ON THE DEVELOPMENT OF MICROBIAL WEED CONTROL AGENTS

No matter how unrestrictive the regulations may be, compliance with regulations will add cost to biological weed controls, especially to microbial herbicides (see Chapter 10). Unlike classical biocontrols, whose development is generally supported through public funds and that are offered to the public free of cost, the microbial herbicides, despite their development through public funds, would be sold to the consumers. The added cost of regulations, which is expected to be borne by the commercial developers of biological control agents, would be passed on to the consumers. One impact of regulations, therefore, could be the higher cost of microbial herbicides and the attractiveness of classical biocontrols to the public. On the other hand, regulations are expected to have a favorable effect in stimulating research and development of new biological weed controls. The negative impact of regulations, such as the higher cost, may be offset by the overall guarantee of safety to the public.

CONCLUSIONS

It is essential to have constructive and flexible regulations that are committed to promoting the use of plant pathogens as biocontrols for weeds. A favorable set of regulatory guidelines are needed to serve as a framework for the development of microbial controls for weeds and to stimulate future research. The proposed and existing regulations represent a positive step in this direction.

REFERENCES

1. R. Charudattan. 1978. *Biological Control Projects in Plant Pathology: A Directory.* Miscellaneous Publication, Plant Pathology Department, University of Florida, Gainesville, FL, 67 pp.
2. J. M. Cullen. 1978. Evaluating the success of the programme for the biological control of *Chondrilla juncea* L. Pp. 117–121 in: T. E. Freeman, ed., *Proceedings of the Fourth*

International Symposium on Biological Control of Weeds. University of Florida, Gainesville, FL.

3. E. Oehrens. 1977. Biological control of the blackberry through the introduction of rust, *Phragmidium violaceum*, in Chile. *FAO Plant Prot. Bull.* **25**, 26–28.

4. L. A. Andres and R. D. Goeden. 1971. The biological control of weeds by introduced natural enemies. Pp. 143–164 in: C. B. Huffaker, ed., *Biological Control*. Plenum Press, New York.

5. B. R. Bartlett, C. P. Clausen, P. DeBach, R. D. Goeden, E. F. Legner, J. A. McMurtry, E. R. Oatmen, E. C. Bay, and D. Rosen. 1978. *Introduced Parasites and Predators of Arthropod Pests and Weeds: A World Review.* Agricultural Handbook No. 480, Agricultural Research Service, U.S. Department of Agriculture, U.S. Government Printing Office, Washington, DC, 545 pp.

6. Anonymous. 1912. Chapter 308, An Act to Regulate the Importation of Nursery Stock and Other Plants and Plant Products. *U.S. Statutes at Large* **37**, 315.

7. Anonymous. 1957. Public Law 85–36, Federal Plant Pest Act. *U.S. Statutes at Large* **71**, 31.

8. D. L. Klingman and J. R. Coulson. 1981. *Guidelines on Proposals to Introduce Foreign Organisms into the United States for the Control of Weeds.* February 26, 1981. Interdepartmental communication, U.S. Department of Agriculture, Science and Education Administration, Working Group on Biological Control of Weeds, Beltsville Agricultural Research Center, Beltsville, MD, 11 pp.

9. R. G. Emge and C. H. Kingsolver. 1977. Biological control of rush skeleton weed with *Puccinia chondrillina*. *Proc. Am. Phytopathol. Soc.* **4**, 215.

10. S. Hasan. 1972. Specificity and host specialization of *Puccinia chondrillina*. *Ann. Appl. Biol.* **72**, 257–263.

11. S. Hasan and A. J. Wapshere. 1973. The biology of *Puccinia chondrillina*, a potential biological control agent of skeleton weed. *Ann. Appl. Biol.* **74**, 325–332.

12. F. J. Simmonds, J. M. Franz, and R. I. Sailer. 1976. History of biological control. Pp. 17–39 in: C. B. Huffaker and P. S. Messenger, eds., *Theory and Practice of Biological Control*. Academic Press, New York.

13. R. Charudattan. 1973. Evaluation of foreign pathogens as biocontrols of hydrilla and waterhyacinth in the U.S.A. In *Proceedings of the Second International Congress of Plant Pathology*. American Phytopathological Society, St. Paul, MN, abstract No. 0390.

14. J. R. Coulson. 1978. U.S. Department of Agriculture, Science and Education Administration, Beltsville Agricultural Research Center-West, Beltsville, MD, personal communication.

15. D. F. Waterhouse. 1980. Commonwealth Scientific and Industrial Research Organization, Division of Entomology, Black Mountain, Canberra, ACT, Australia, personal communication.

16. G. Mathys. 1977. Society supported disease management activities. Pp. 363–380 in: J. G. Horsfall and E. B. Cowling, eds., *Plant Disease: An Advanced Treatise*. Vol. 1. *How Disease is Managed*. Academic Press, New York.

17. Anonymous. 1980. Chapter 6, Insecticides and Environmental Pesticide Control. *United States Code Annotated*. **Title 7**, 244.

18. Anonymous. 1980. *Guidelines for Registering Pesticides in the United States.* Subpart M. *Data Requirement for Biorational Pesticides. Preamble and Guidelines.* September 29, 1980. The Environmental Protection Agency, Hazard Evaluation Division, Office of Pesticide Programs, Washington, DC, 433 pp.

19. Anonymous. 1981. Certain biological control agents: Proposed exemption from regulation. *Fed. Regist.* **46** (56), 18322–18325.

20. M. Rogoff. 1980. U.S. Department of Agriculture, Science and Education Administration, Beltsville Agricultural Research Center-West, Beltsville, MD, personal communication.

21. B. Lundholm and M. Stackerud. 1980. *Environmental Protection and Biological Forms of Control of Pest Organisms.* Ecological Bulletins 31, Swedish Natural Science Research Council, Box 23136, S-104 35 Stockholm, Sweden, 171 pp.

12

INTEGRATION OF MICROBIAL HERBICIDES WITH EXISTING PEST MANAGEMENT PROGRAMS

ROY J. SMITH, JR.

Agricultural Research, Science and Education Administration,
U.S. Department Agriculture, Stuttgart, Arkansas

Pest control is an acceptable and necessary component of modern agriculture and is required for the protection of public health and welfare (1,2). Some pest control methods used in the past, however, have produced undesirable side effects. With the growing emphasis on integrated pest management systems, needs for new types of pest control technology including biological control methods can be expected to increase in the future. Biological agents such as pathogens and insects have been used to control weeds, and the basic principles of the operational use of these agents are reasonably well understood (2). Considerable research will be required, however, before biological weed control agents can be used judiciously in integrated weed and pest management systems.

Cooperative contribution of Agricultural Research, Science and Education Administration, U.S. Department of Agriculture and the University of Arkansas Agricultural Experiment Station, Stuttgart, Arkansas.

189

INTEGRATED PEST MANAGEMENT SYSTEMS

In integrated pest management systems the principles, practices, materials, and strategies of control are used efficiently to control pests while minimizing undesirable effects (3–5). Integrated pest management systems require several pest control components to deal with the diversity of pest problems. Integrated pest management systems in crops include (1) the integration of several methods to control a single pest, (2) the integration of many methods against a complex of pests infesting a single crop, (3) the integration of many methods against a complex of pests attacking several crops, (4) the integration of pest management technology with other crop management systems, and (5) the integration of pest management systems with agroecosystems on farms and in areas and regions.

Effective integrated pest management systems must be part of the overall crop management system (3–5). A thorough understanding of the complex crop management operations can be obtained through the systems approach. This approach takes into account the need to increase agricultural production and to determine economic losses, risks to human health and safety, potential for damage from pesticides to nontarget organisms, environmental quality, and energy requirements. Among the highly desirable pest control components available are the use of selective pesticides; pathogens, predators, and parasites that attack pests; genetic and sterility methods to suppress reproduction of pests; and crop cultivars that are resistant to several pests (3). The systems approach also includes the use of ecological, biological, cultural, chemical, and mechanical practices for controlling pests in crops; habitat management; protection of wildlife; biochemical attractants that divert and entrap pests; and pest repellents.

INTEGRATED WEED MANAGEMENT SYSTEMS

Integrated weed management systems are viable components of integrated pest management systems (3). They emphasize a specifically directed agroecosystem approach for the management and control of weed populations. The weed management system combines the use of (1) multiple-pest-resistant, high-yielding, well-adapted crop cultivars that also resist weed competition and (2) a precise placement and timing of fertilizers to give the crop a competitive advantage. Seedbed tillage and seeding methods that enhance crop growth while minimizing weed growth, optimum crop plant populations, and the use of crop cultivars that form a canopy for shading early season weed growth are parts of the weed management system. Such systems also include the use of judicious irrigation practices; timely and appropriate cultivations; carefully planned crop rotations; field sanitation; harvesting methods that do not spread weed seeds; use of biological control agents such as pathogens, insects, and nematodes; and effective chemical weed control methods. For the systems-directed agroecosystem approach to be most effective, preventive weed control practices must precede and accompany the integrated weed management system to reduce the numbers of weed seeds and other propagules in the soil.

BIOLOGICAL CONTROL OF WEEDS

Biological weed control agents are organisms that suppress or kill weedy plants without significantly injuring desirable plants (6). They include insects, pathogens, nematodes, parasitic plants, and competing plants.

The great diversity of insects throughout the world makes them prime candidates for research and development as biocontrol agents for weeds. Several phytophagous insects have been used to control weeds (6), and successful examples include the use of the moth *Cactoblastis cactorum* for control of pricklypear in Australia; the beetle *Chrysolina quadrigemina* for control of St. Johnswort in North America; and a combination of a leaf-feeding beetle, *Agasicles hygrophila*, and a stem boring moth, *Vogtia malloi*, for control of alligatorweed in the southern United States (6).

There are numerous other projects aimed at developing insects as biocontrol agents for weeds (7,8). Most of the attempts to use insects have been for the control of introduced weeds because of the general absence of natural enemies in the weeds' new environment (2). Recent attempts to control weeds by insects include the use of a native moth, *Bactra verutana*, for control of purple nutsedge (9–11); the spurge hawkmoth *Hyles euphorbiae* for control of leafy spurge (12); the mullein moth *Cucullia verbasci* for control of common mullein (13); and a weevil, *Rhinocyllus conicus*, for control of milk and musk thistles (14–16).

The ability to control weeds in agroecosystems with plant pathogens has been amply demonstrated (see Chapter 3). The use of pathogens for weed control has been separated into the classical and the bioherbicide tactics (see Chapter 3 and Ref. 17).

Successful examples of pathogens used in the classical tactic include two rust fungi, *Puccinia chondrillina* for control of rush skeletonweed in Australia and in the western United States, and *Uromyces rumicis* for suppression of curly dock in pastures in Europe (see Refs. 2 and 17 and Chapter 3 for other literature). Successful examples of pathogens applied by the microbial herbicide tactic include three anthracnose-inducing pathogens:

Colletotrichum gloeosporioides f. sp. *aeschynomene* (CGA) for control of northern jointvetch in rice and soybean (18,19), *C. gloeosporioides* f. sp. *jussiaeae* (CGJ) for control of winged waterprimrose in rice (19), and *C. malvarum* (CM) for control of prickly sida in soybean and cotton (20); a soil-borne fungus, *Phytophthora palmivora* (formerly considered to be *P. citrophthora*) for control of strangler vine in citrus (21); a leaf spot-inducing fungus, *Cercospora rodmanii*, for suppression of waterhyacinth (22); and a blight-inducing fungus, *Alternaria macrospora* (AM), for control of spurred anoda (23).

INTEGRATION OF MICROBIAL HERBICIDES WITH PEST MANAGEMENT PROGRAMS

Integration of plant pathogens with other control practices such as the use of insect biocontrols and herbicides would improve the prospects for weed control over the use of only one such control practice. For example, insect-damaged plants of waterhyacinth were more severely infected with plant pathogens than were plants not damaged by insects (24). Waterhyacinth plants damaged by the feeding of two arthropods (*Neochetina eichhorniae*, a weevil, and *Orthogalumna terebrantis*, a mite) were more frequently infected by a virulent pathogen, *Acremonium zonatum*, and other fungi and bacteria than were insect-free plants (24).

Biological control agents including microbial herbicides may be integrated into weed and pest management systems employed by the farmer in a crop production system. There is limited research and technology, however, on the use of microbial herbicides in a total pest-managed, crop production system. *Phytophthora palmivora*, a microbial herbicide used to control strangler vine in citrus in Florida, was inhibited by some herbicide treatments (25). In these tests, germinability of chlamydospores of the fungus was reduced when tank-mixed with herbicides such as diuron, glyphosate, and paraquat. However, sequential application of spore suspensions and herbicides prevented the adverse effects of the herbicide on the fungus. In this research, spore germination was not reduced when water suspensions of chlamydospores were applied 3 weeks after spraying glyphosate. Thus both glyphosate and the microbial herbicide were active and controlled the weed effectively.

The use of endemic fungi applied as microbial herbicides for weed control in agronomic crops offers some unusual opportunities for integration with weed and pest management systems. However, for successful use of endemic fungi as biological weed control agents in such systems, we must consider the pathogen, the weed, the environment, the influence that these fungi have on the pest management and crop production systems, and the influence of these systems on the fungi. I shall discuss the use of microbial herbicides as components of integrated weed and pest management systems for three agronomic crops—rice, soybean, and cotton.

INTEGRATION OF BIOCONTROL AGENTS WITH A PEST MANAGEMENT PROGRAM FOR RICE

Integration of all available weed control technologies is required to reduce losses in yield and quality of rice and to minimize the potential for environmental damage (26). The best approach to weed control in rice is through an integrated weed management system that combines preventive, cultural, mechanical, chemical, and biological control practices. The system that omits any one of these components is usually inadequate (27). Furthermore, the weed control program must be integrated with other plant protection technologies. Particularly, the disease and insect control practices used must not be incompatible with the specific weed control practice or vice versa.

PATHOGENS FOR PROBLEM WEEDS

Three fungal pathogens have been used to control three problem weeds in rice (17–20). These agents, CGA, CGJ, and CM, are specific, endemic fungal pathogens that infect northern jointvetch, winged waterprimrose, and prickly sida, respectively. Any one or combination of these three weeds may be economically important in rice fields. Northern jointvetch and winged waterprimrose may infest the flooded paddy and the levees, but prickly sida, which does not tolerate flooding, usually infests the levees only. Depending on the weed problem, the three fungi may be used alone or in combination for control of one or any combination of these three weeds.

HERBICIDES FOR PROBLEM WEEDS

More than 50 weed species infest direct-seeded rice in the United States (27). The use of chemical herbicides is an important component of an effective weed control program for rice, and several herbicides are registered for use in rice (Table 1). These herbicides used alone or in combination control most of the problem weeds in rice. Although each one is efficacious on more than one weed species, each herbicide may have some inadequacies. The prominent inadequacies include (1) injury to nontarget crops as

TABLE 1. Selected Pesticides Used in Three Crop Production Systems Grouped by Crop and Pesticide Type[a]

Crop	Herbicide[b]	Fungicide[c]	Insecticide[d]	Nematicide[c]
Cotton	Cyanazine	Captan	Acephate	Aldicarb
	Dalapon	Carboxin	Aldicarb	Phenamiphos
	Dinoseb	Chloroneb	Azinphos-methyl	
	Diuron	Maneb	Carbaryl	
	DSMA	PCNB	Chlordimeform	
	Fluchloralin	PCNB-ETMT	Chlorpyrifos	
	Fluometuron	TCMTB	Dicrotophos	
	Glyphosate	Thiram	Dimethoate	
	Linuron		Disulfoton	
	MSMA		Endosulfan	
	Norflurazon		Endrin	
	Paraquat		EPN	
	Pendimethalin		Fenvalerate	
	Profluralin		Malathion	
	Prometryn		Methamidophos	
	Trifluralin		Methidathion	
			Methomyl	
			Monocrotophos	
			Permethrin	
			Phorate	
			Propargite	
			Sulprofos	
			Toxaphene	
			Trichlorfon	
Soybean	Alachlor	Benomyl	Acephate	Aldicarb
	Bentazon	Captan	Carbaryl	D-D mixture
	Dalapon	Carboxin	Methomyl	Ethoprop
	2,4-DB	PCNB-ETMT	Methyl parathion	Metham
	Dinoseb	Thiabendazole	Toxaphene	Phenamiphos
	Fluchloralin	Thiram		
	Glyphosate			
	Linuron			
	Metolachlor			
	Metribuzin			
	Naptalam			
	Oryzalin			
	Paraquat			

TABLE 1. *(Continued)*

Crop	Pesticide Type			
	Herbicide[b]	Fungicide[c]	Insecticide[d]	Nematicide[c]
	Pendimethalin			
	Profluralin			
	Trifluralin			
	Vernolate			
Rice	Bentazon	Benomyl	Bufencarb	None
	Bifenox	Captafol	Carbaryl	
	Butachlor	Captan	Carbofuran	
	2,4-D	Carboxin	Malathion	
	MCPA	PCNB-ETMT	Methyl parathion	
	Molinate	TCMTB	Toxaphene	
	Oxadiazon	Thiram		
	Propanil	Zinc ion-maneb		
	Silvex	complex		
	2,4,5-T			
	Thiobencarb			

[a] See Appendix 4 for chemical names of these pesticides.
[b] *Source*:   Refs. 27 and 39.
[c] *Source*:   Ref. 40.
[d] *Source*:   Ref. 41.

a result of spray drift or residues in soil, (2) injury to the treated crop through improper or untimely application or through adverse interactions with other chemical pesticides, (3) lack of residual efficacy, thus requiring retreatment, (4) a narrow spectrum of activity, especially when there is a close taxonomic relationship between the weed and the crop, and (5) alteration of environmental quality beyond tolerable limits.

FUNGICIDES AND INSECTICIDES FOR RICE DISEASES AND INSECTS

Several economically important diseases and insects occur on rice in the United States (28,29). The use of fungicides, insecticides, and other crop management practices is an essential component of a pest management program in rice. Seedling and foliar diseases are controlled by the application of fungicides as preplant seed treatments or as foliar sprays at midseason (Table 1). Several insecticides (Table 1) are required for control of insects such as rice water weevils (*Lissorhoptrus oryzophilus*), chinch bugs (*Blissus leucopterus*), fall armyworms (*Spodoptera frugiperda*), and rice stink bugs (*Oebalus pugnax*).

Although these fungicides and insecticides control rice diseases and insects effectively, some of them may interact adversely with other pest management practices. For example, carbamate and organophosphate insecticides inhibit an enzyme in the rice plant that detoxifies propanil (see Appendix 4 for chemical names of pesticides cited in this book), resulting in phytotoxicity to rice from propanil (27). These classes of insecticide cannot be applied safely within 14 days before or after application of propanil. Chlorinated hydrocarbon insecticides that do not affect this enzyme or interact adversely with propanil may be substituted for carbamate or organophosphate insecticides for insect control in rice.

INTEGRATION OF MICROBIAL HERBICIDES WITH WEED CONTROL PROGRAMS

A principal deterrent to the commercial use of a pathogen is its specificity and control of only one weed species (see Chapter 10). A microbial herbicide may be integrated with other biological agents, chemical herbicides, or other practices to control the complex of weed flora in rice fields. Also, two or more microbial herbicides may be combined as tank mixtures or used sequentially to control several weed species. Such use of more than one pathogen in tank mixtures may help overcome the problem of specificity.

Tank mixtures of two fungi have successfully controlled two weeds in rice fields. Northern jointvetch and winged waterprimrose were controlled by applying a tank mixture of CGA and CGJ. In a field test 100 % of the winged waterprimrose and 96 % of the northern jointvetch plants were killed 4 weeks after aerial application with a tank mixture of these two fungi at midseason (19). A tank mixture consisting of three fungi was applied aerially to a rice field at Stuttgart, Arkansas in 1980 (30). In this test CGA and CGJ controlled northern jointvetch and winged waterprimrose, but CM failed to control prickly sida. The unusually high temperatures (35°–42°C) during late June and July of that year prevented the development of CM on prickly sida. In laboratory studies the growth and activity of CM are reduced significantly by temperatures above 30°C (20). Early-season (May or early June) application of CM before field temperatures become too high for its activity may overcome reduced efficacy at high temperatures. Subsequently, a tank mixture of CGA and

CGJ could be applied at midseason for control of northern jointvetch and winged waterprimrose. The efficacies of these two pathogens have been unaffected by high field temperatures.

Although tank mixtures of pathogenic fungi and chemical herbicides have been unsuccessful, sequential applications of microbial herbicides (CGA and CGJ) with chemical herbicides have controlled a broad spectrum of weeds. For example, some herbicides such as propanil, molinate, and (2,4,5-trichlorophenoxy)acetic acid (2,4,5-T) that were combined with spore suspensions of CGA inhibited spore germination and vegetative growth (31). Therefore, tank mixtures of these chemical herbicides and CGA or CGJ were not compatible and could not be employed together to increase weed control. Research has also shown that tank mixtures of propanil and bentazon applied early in the growing season controlled barnyardgrass, spreading dayflower, ducksalad, redstem, and waterhyssop; midseason applications of a tank mixture of CGA and CGJ controlled northern jointvetch and suppressed winged waterprimrose (32). All northern jointvetch plants were killed, but winged waterprimrose plants were only partially controlled (32). These results indicated that the winged waterprimrose plants were too large at the time of CGJ application for its optimum pathogenicity and that effective control of this weed could have been obtained with more timely applications of the microbial herbicide.

INTEGRATION OF MICROBIAL HERBICIDES WITH A PEST MANAGEMENT SYSTEM

For successful integration with a pest management system that uses fungicides, insecticides, and nematicides, the microbial herbicides must not interact adversely with the disease, insect, or nematode control programs practiced. Single or multiple microbial herbicides may be applied in tank mixtures or sprayed sequentially with fungicides, insecticides, and nematicides for controlling the pest complex of weeds, diseases, insects, and nematodes.

The microbial herbicide containing CGA has been used with certain fungicides and insecticides to control a weed, fungal diseases, and insects in rice. Spore suspensions of CGA applied at midseason just as the northern jointvetch plants emerged through the rice canopy controlled all the northern jointvetch weeds after carbofuran or the experimental insecticide isofenphos had been applied during the early season for control of rice water weevils (32). Also, benomyl or the experimental fungicide tricyclazole did not affect the performance of CGA when applied 2–3 weeks after applying the pathogen. In this experiment the rice field was drained for a 4-day period about 2 weeks before spraying CGA to permit soil drying to control "straighthead," a physiological disorder.

Adverse interactions of CGA with fungicides have been observed. For example, CGA failed to control northern jointvetch effectively when spore suspensions were unintentionally combined with benomyl prior to an aerial application (33). Also, in experiments designed to determine the competition between rice and northern jointvetch, benomyl successfully controlled natural CGA infections on the weed (34). Therefore, research indicates that benomyl applied in combination as tank mixtures or sequentially with CGA may indeed affect disease development and reduce subsequent control of northern jointvetch plants.

However, when managed in a judicious pest control program, CGA controls northern jointvetch. Precise timing of applications of spores and the incompatible fungicides and insecticides is critical for disease development to occur on the weed. Tank mixtures of CGA and CGJ applied at the optimum time (when northern jointvetch and winged waterprimrose were emerging through the crop canopy) controlled these weeds even though conventional treatments of benomyl were applied subsequently to control rice diseases (35).

INTEGRATION OF MICROBIAL HERBICIDES WITH PEST MANAGEMENT PROGRAMS IN TWO OTHER CROP PRODUCTION SYSTEMS

INTEGRATION OF MICROBIAL HERBICIDES WITH A RICE-SOYBEAN CROPPING SYSTEM

Colletotrichum gloeosporioides f. sp. *aeschynomene* controls northern jointvetch in rice and soybean (36), CGJ controls winged waterprimrose in rice (19), and CM controls prickly sida in soybean (20). During the rice production year CGA and CGJ may be applied in tank mixtures or sequentially to control northern jointvetch and winged waterprimrose. Likewise, during the soybean production year CGA and CM may be applied in combination or sequentially to control northern jointvetch and prickly sida. Consequently, these three fungi may be used alone or in combinations for control of these three weeds in rice and soybean.

Numerous pesticides are used for control of weeds, diseases, insects, and nematodes in rice and soybean (Table 1). Some of these pesticides interact adversely with the pathogenic fungi used for weed control. For example, benomyl, chlorothalonil, and thiabendazole reduced CGA infection on northern jointvetch plants when applied 3 days before and as soon as 1 hour after CGA inoculations (37). Benomyl and thiabendazole reduced CGA infections when applied 2 days after the fungus, but chlorothalonil did not when applied at a similar interval. Successful control of northern jointvetch in soybean fields with CGA has been achieved by delaying benomyl application until after the weeds had developed severe disease symptoms, usually 2–3 weeks after application of the fungus. Therefore, as in the rice cropping system discussed earlier, the use of fungicides for control of diseases in rice and soybean should be manipulated so as not to interfere with the activity of CGA on northern jointvetch.

Most of the pesticides listed in Table 1 have not been researched to determine their influence on the pathogenicity of CGA and CGJ and their subsequent weed control potential.

In an integrated pest management system, factors other than pest control must be considered. For example, crop management is an important component of the system. For CGA to infect and kill northern jointvetch in soybean, the soil must be saturated with water at the time of applying spores of the fungus (36). Therefore, applications of CGA can be timed to coincide with rain or irrigation that saturates the soil for a microenvironment favorable for disease development.

INTEGRATION OF MICROBIAL HERBICIDES WITH A SOYBEAN-COTTON CROPPING SYSTEM

In separate studies CGA and CM were used to control northern jointvetch and prickly sida, respectively, in soybean (20,36), and CM was used for control of prickly sida in cotton (20). Preemergence or postemergence applications of AM control spurred anoda in cotton (38). During the soybean production year CGA and CM may be applied in tank mixtures or sequentially for control of northern jointvetch and prickly sida. Likewise, in the cotton production year CM and AM may be applied for control of prickly sida and spurred anoda. Consequently, these three fungi may be used alone or in combination for control of the three weeds in the two crops.

Numerous pesticides are used for control of weeds, diseases, insects, and nematodes in soybean and cotton (Table 1). As stated previously, the application of benomyl before or just after the CGA treatment reduces the activity of this pathogen on northern jointvetch. The influence of other pesticides used in a soybean-cotton cropping system on the weed control activity of CGA, CM, and AM is unknown. Further research on this aspect is needed.

In a soybean-cotton cropping system, microbial herbicides may be integrated with other biological weed control agents as well as with pesticides. For example, CM and AM may be incorporated into a weed control program for cotton that includes the use of *Bactra verutana* for control of purple nutsedge. In such a weed management system, insecticides that would not injure *Bactra* would have to be selected for general insect control programs.

CONCLUSIONS

Integration of all weed and pest control technologies is required for reduction of losses in yield and quality of crops caused by weeds, insects, diseases, and nematodes and for minimization of the potential for environmental damage. Effective weed and pest management systems integrate the use of preventive practices; crop rotation; soil and water management practices; cultivation; crop management; biological control agents such as pathogens, insects, and nematodes; chemical herbicides; and other pesticides. The use of weed pathogens, together with other control practices aimed at reducing losses by weeds, permits the farmer to abandon some of the more traditional or inefficient agronomic practices. For example, the integrated use of microbial and chemical herbicides may allow a more efficient use of irrigation water, fertilizer, and energy. The use of microbial herbicides in integrated weed management systems must be compatible with other practices for managing pests, increasing production, and ensuring a quality environment.

Crops will be confronted with shifts in weed flora as susceptible weeds are removed and are replaced by more tolerant weeds in the crop monoculture as a result of the continued use of chemical herbicides. New approaches, including biological control, will be required for management of new weed problems. Research has demonstrated that weed pathogenic fungi can be used along with standard chemical herbicides and other weed control practices to control a spectrum of weed species. Presently, several weed pathogenic fungi are available that show promise for control of weeds that are not now controlled by conventional treatments.

Integration of microbial herbicides as viable components of weed management systems will be a challenge to researchers and organizations interested in pest management sciences. Costs, benefits, and risks of all components of integrated weed and pest management systems must be examined carefully. Weed pathogens offer unusual opportunities for the development of weed control practices that will be compatible with other components of integrated pest management systems.

REFERENCES

1. J. D. Fryer. 1977. Introduction. Pp. xi–xiv in: J. D. Fryer and S. Matsunaka, eds., *Integrated Control of Weeds*. University of Tokyo Press, Tokyo, Japan.

2. Anonymous. 1978. *Biological Agents for Pest Control—Status and Prospects*. U.S. Department of Agriculture, U.S. Government Printing Office, Washington, DC, 138 pp.

3. W. C. Shaw. 1979. Integrated weed management systems technology. Pp. 149–157 in: J. M. Brown, ed., *Proceedings of the Beltwide Cotton Production Research Conferences*. National Cotton Council, Memphis, TN.

4. W. B. Ennis, Jr. 1974. Weed science in pest management programs. *Proc. South. Weed Sci. Soc.* 27, 8–15.

5. W. B. Ennis, Jr. 1977. Integration of weed control technologies. Pp. 229–242 in: J. D. Fryer and S. Matsunaka, eds., *Integrated Control of Weeds*. University of Tokyo Press, Tokyo, Japan.

6. L. A. Andres. 1977. The biological control of weeds. Pp. 153–174 in: J. D. Fryer and S. Matsunaka, eds., *Integrated Control of Weeds*. University of Tokyo Press, Tokyo, Japan.

7. R. D. Goeden, L. A. Andres, T. E. Freeman, P. Harris, R. L. Pienkowski, and C. R. Walker. 1974. Present status of projects on the biological control of weeds with insects and plant pathogens in the United States and Canada. *Weed Sci.* 22, 490–495.

8. P. Harris. 1979. Cost of biological control of weeds by insects in Canada. *Weed Sci.* 27, 242–250.

9. K. E. Frick and P. C. Quimby, Jr. 1977. Biocontrol of purple nutsedge by *Bactra verutana* Zeller in a greenhouse. *Weed Sci.* 25, 13–17.

10. K. E. Frick, R. D. Williams, and R. F. Wilson. 1978. Interactions between *Bactra verutana* and the development of purple nutsedge (*Cyperus rotundus*) grown under three temperature regimes. *Weed Sci.* 26, 550–553.

11. K. E. Frick, R. D. Williams, P. C. Quimby, Jr., and R. F. Wilson. 1979. Comparative biocontrol of purple nutsedge (*Cyperus rotundus*) and yellow nutsedge (*C. esculentus*) with *Bactra verutana* under greenhouse conditions. *Weed Sci.* 27, 178–183.

12. J. R. Forwood and M. K. McCarty. 1980. Control of leafy spurge (*Euphorbia esula*) in Nebraska with the spurge hawkmoth (*Hyles euphorbiae*). *Weed Sci.* 28, 235–240.

13. M. G. Maw. 1980. *Cucullia verbasci* an agent for the biological control of common mullein (*Verbascum thapsus*). *Weed Sci.* 28, 27–30.

14. J. M. Hodgson and N. E. Rees. 1976. Dispersal of *Rhinocyllus conicus* for biocontrol of musk thistle. *Weed Sci.* 24, 59–62.

15. R. D. Goeden and D. W. Ricker. 1977. Establishment of *Rhinocyllus conicus* on milk thistle in southern California. *Weed Sci.* 25, 288–292.

16. B. Puttler, S. H. Long, and E. J. Peters. 1978. Establishment in Missouri of *Rhinocyllus conicus* for the biological control of musk thistle (*Carduus nutans*). *Weed Sci.* 26, 188–190.

17. G. E. Templeton and R. J. Smith, Jr. 1977. Managing weeds with pathogens. Pp. 167–176 in: J. G. Horsfall and E. B. Cowling, eds., *Plant Disease: An Advance Treatise*. Vol. 1. *How Disease is Managed*. Academic Press, New York.

18. J. T. Daniel, G. E. Templeton, R. J. Smith, Jr., and W. T. Fox. 1973. Biological control of northern jointvetch in rice with an endemic fungal disease. *Weed Sci.* 21, 303–307.

19. C. D. Boyette, G. E. Templeton, and R. J. Smith, Jr. 1979. Control of winged water-primrose (*Jussiaea decurrens*) and northern jointvetch (*Aeschynomene virginica*) with fungal pathogens. *Weed Sci.* **27**, 497–501.

20. T. L. Kirkpatrick. 1978. Bioherbicidal potential of *Colletotrichum malvarum* for prickly sida control. M. S. thesis. University of Arkansas, Fayetteville, AR, 47 pp.

21. H. C. Burnett, D. P. H. Tucker, and W. H. Ridings. 1974. Phytophthora root and stem rot of milkweed vine. *Plant Dis. Rep.* **58**, 355–357.

22. K. E. Conway, T. E. Freeman, and R. Charudattan. 1978. Development of *Cercospora rodmanii* as a biological control for *Eichhornia crassipes*. Pp. 225–230 in: *Proceedings of the European Weed Research Society Fifth Symposium on Aquatic Weeds.* P. O. Box 14, Wageningen, The Netherlands.

23. H. L. Walker and G. L. Sciumbato. 1979. Evaluation of *Alternaria macrospora* as a potential biocontrol agent for spurred anoda (*Anoda cristata*): Host range studies. *Weed Sci.* **27**, 612–614.

24. R. Charudattan, B. D. Perkins, and R. C. Littell. 1978. Effects of fungi and bacteria on the decline of arthropod-damaged waterhyacinth (*Eichornia crassipes*) in Florida. *Weed Sci.* **26**, 101–107.

25. M. E. Kannwischer. 1980. Plant Pathology Department, University of Florida, Gainesville, FL, personal communication.

26. R. J. Smith, Jr. and K. Moody. 1979. Weed control practices in rice. Abstracts of Papers of the Ninth International Congress of Plant Protection and 71st Annual Meeting of the American Phytopathological Society, August 5–11, 1979. Washington, DC, Abstract No. 267.

27. R. J. Smith, Jr., W. T. Flinchum, and D. E. Seaman. 1977. *Weed Control in U.S. Rice Production.* Agricultural Handbook No. 497. U.S. Department of Agriculture, U.S. Government Printing Office, Washington, DC, 78 pp.

28. J. G. Atkins. 1973. Rice diseases. Pp. 141–150 in: Anonymous, ed., *Rice in the United States: Varieties and Production.* Agricultural Handbook No. 289. U.S. Department of Agriculture, U.S. Government Printing Office, Washington, DC.

29. J. R. Gifford. 1973. Insects and their control. Pp. 151–154 in: Anonymous, ed., *Rice in the United States: Varieties and Production.* Agricultural Handbook No. 289. U.S. Department of Agriculture, U.S. Government Printing Office, Washington, DC.

30. R. J. Smith, Jr. 1980. Unpublished data.

31. D. O. TeBeest and G. E. Templeton. 1975. Department of Plant Pathology, University of Arkansas, Fayetteville, AR, personal communication.

32. M. A. Colbert. 1979. Department of Plant Pathology, University of Arkansas, Fayetteville, AR, personal communication.

33. D. O. TeBeest. 1976. Department of Plant Pathology, University of Arkansas, Fayetteville, AR, personal communication.

34. R. J. Smith, Jr. 1971. Unpublished data.

35. R. J. Smith, Jr. 1979. Unpublished data.

36. R. J. Smith, Jr., G. E. Templeton, and D. O. TeBeest. 1978. Field efficacy of a fungus for control of northern jointvetch in a 3-year pilot test. *Proc. Rice Tech. Work. Group* **17**, 71–72.

37. D. O. TeBeest. 1980. Department of Plant Pathology, University of Arkansas, Fayetteville, AR, personal communication.

38. H. L. Walker. 1980. USDA-SEA-AR, Southern Weed Science Laboratory, P. O. Box 225, Stoneville, MS, personal communication.

39. Anonymous. 1980. *Recommended Chemicals for Weed and Brush Control.* Miscellaneous Publication No. 44. Cooperative Extension Service, University of Arkansas, Fayetteville, AR, 76 pp.

40. Anonymous. 1980. *Abstract of Pesticides Suggested for Plant Disease Control.* Miscellaneous Publication No. 154. Cooperative Extension Service, University of Arkansas, Fayetteville, AR, 18 pp.

41. Anonymous. 1980. *Insecticide Recommendations for Arkansas.* Miscellaneous Publication No. 144. Cooperative Extension Service, University of Arkansas, Fayetteville, AR, 96 pp.

IV

SUMMARY

13

LITERATURE RETRIEVAL FOR THE EVALUATION OF BIOCONTROL AGENTS

RICHARD T. HANLIN

*Department of Plant Pathology, University of Georgia,
Athens, Georgia*

As in any area of scientific research, literature retrieval is an important aspect of studies on the biological control of weeds. In one sense the problems of searching the literature for pertinent references on this subject are compounded because two groups of organisms are involved, namely, the host weed and the pathogen, and the sources of information for each may be quite different.

ABSTRACTING JOURNALS

The three primary abstracting journals of relevance to biological weed control are *Abstracts of Mycology* (AM), *Review of Plant Pathology* (RPP), and *Weed Abstracts* (WA), all of which provide complete coverage in printed form for manual retrieval. *Abstracts of Mycology* is compiled from *Biological Abstracts* (BA) (1) and RPP and WA are prepared by the Commonwealth Agricultural Bureaux (CAB) in England (2). Because of the interdisciplinary nature of weed biocontrol research, however, it probably would be advantageous to use a broader data base to ensure that no pertinent references are overlooked. The new CAB journal, *Biocontrol News and Information* (3), should be especially useful for this purpose. Although abstracting journals are available in printed form, computer searching can be utilized to provide speedy retrieval of references. Another advantage of computer searches is that data bases are often issued before the printed copy becomes available. The main handicap to computer searching, however, is that few data bases extend prior to 1970.

COMPUTERIZED ABSTRACTING SERVICES

The two most widely used computerized abstracting services applicable to agriculture are the BioSciences Information Service (BIOSIS) and the Agricultural On-Line Access [AGRICOLA, formerly known as Cataloging and Indexing (CAIN)] systems. The BIOSIS service (4) is prepared by a commercial firm, and AGRICOLA is prepared by the National Agricultural Library (5) of the U.S. Department of Agriculture (USDA). Within BIOSIS (6) there are two data bases of primary interest, Biological Abstracts (BA) and Biological Abstracts/Reports, Reviews, Meetings (BA/RRM), formerly known as Bioresearch Index (BIOI). These two data bases, referred to collectively as Biosis Previews (7), differ in coverage. *Biological Abstracts* includes new books, letters and notes, and research journals; provides coverage of 9000 journals from 100 countries; and publishes some 150,000 citations annually in biweekly issues. The BA/RRM service covers annual reports, bibliographies, book chapters, data reports, letters, notes, review journals, research journals, translations of Russian journals, symposium abstracts, and symposium papers. In 1979 BA/RRM published approximately 125,000 citations, and BIOSIS reviews now contain over 2.25 million citations.

The AGRICOLA system provides coverage of the *Bibliography of Agriculture*, the *American Bibliography of Agricultural Economics*, the *American Agricultural Economics Documentation Center* (AAEDC), the *Food and Nutrition Information Center* (FNIC), the *Pesticides Documentation Bulletin*, and *Agriculture Canada* (8). Some 15,000–18,000 citations are published monthly. The AGRICOLA data base currently contains over 1.25 million citations.

Each of these data bases can be searched by subject, author, concept area, taxonomic category (*Biosystematic Index*), or any combination of these (9). One can, for example, retrieve all the references on northern jointvetch (*Aeschynomene virginica*) by either common or scientific name or both; the same can be done for the anthracnose pathogen, *Colletotrichum gloeosporioides* f. sp. *aeschynomene.* Since this might result in references not relevant to the study, a likely profile would search for references that include northern jointvetch as well as anthracnose; this would retrieve only those references in which both organisms were discussed.

The Commonwealth Agricultural Bureaux (CAB) also provides a computerized version of citations from its 45 abstracting journals, including *Biocontrol News and Information, Review of Plant Pathology* (formerly *Review of Applied Mycology*), and *Weed Abstracts*. The computerized version is issued monthly and covers over 8000 journals (10). The CAB data base contains approximately 900,000 citations.

Workers in United States federal and land grant institutions also have access to the USDA Current Research Information Service (CRIS) and the Current Awareness Literature Service (CALS) data bases; CRIS contains project titles and summaries for all current USDA and State Experiment Station projects, and CALS provides a searching system that covers BA, BA/RRM, *Chemical Abstracts, Engineering Index, Food Science Technology Abstracts, Government Reports Announcement* of the National Technical Information Service (NTIS), and the CAB data file. These data bases are designed to keep researchers up-to-date in their areas of interest.

There are two additional services that may be purchased. The Smithsonian Science Information Exchange (SSIE) is a nonprofit organization that catalogs projects sponsored by federal, state, and local government agencies, nonprofit associations and foundations, colleges and universities, and foreign research organizations. Project information is contained in the SSIE *Notice of Research Project* (NRP), which gives a summary of the work under way. Information packages consisting of NRPs in specific areas such as weed control are available for a fee. Over a dozen such packages are available on specific weeds (11).

The DIALOG system is an on-line information service that is available from Lockheed Information Systems (12) for a fee. In 1978, 94 data bases covering nearly every subject were available for searching. Printed copies can also be purchased from the DIALOG system.

HERBARIUM RECORDS

In the area of biocontrol of weeds with pathogens there is another source of information that may be useful to the researcher. This is the information contained in sources such as herbarium and plant disease clinic records. Although these records are seldom referred to when considering information retrieval, they contain a wealth of information that is available to the investigator. Several questions arise when considering the possibility of controlling a weed with a pathogen. What is the distribution of the weed, and is this distribution coincident with that of the crop involved? Does the pathogen have the same distribution as the host? If not, is this due to geographic or climatic factors, or does it merely reflect the lack of data? What are the natural hosts for the pathogen, and in the case of an exotic weed, is there an effective pathogen in the native habitat that could be imported for use? Field observations are essential in any such studies, but individual scientists seldom can cover large areas in person. Valuable information can be obtained from herbarium and plant disease clinic records, but unfortunately, retrieving data from such records can be a slow and arduous process.

Herbaria, of course, can serve as repositories for voucher specimens of both weed and pathogen (Figs. 1–3). Herbarium records can also provide data on the distribution of the weed host as a supplement to field observations. Information on related species that may serve as possible hosts for the pathogen also may be available from herbarium records. Obviously, the amount of data that can be obtained from this source depends on the quality and size of the herbarium collection. Checklists such as the one for Georgia based on specimens in the University of Georgia herbarium (13) are especially useful. In a few instances regional weed manuals that contain information on distribution are also available (14). Examination of

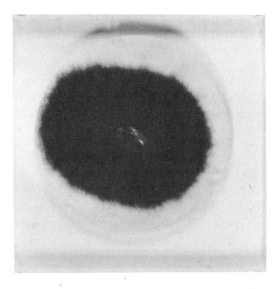

FIGURE 1. A freeze-dried culture of *Exserohilum rostratum* preserved as a specimen in the Julian H. Miller Mycological Herbarium, University of Georgia.

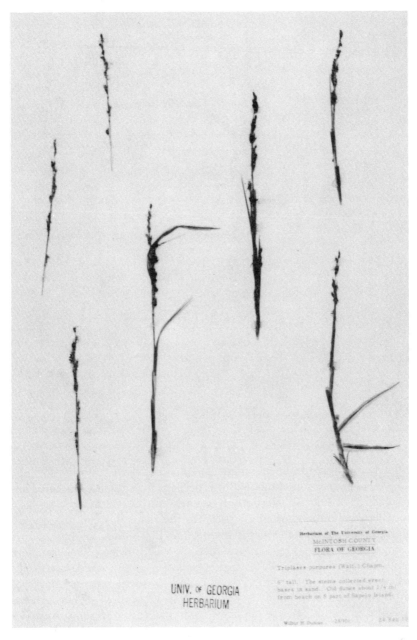

Text within the image:

UNIV. OF GEORGIA
HERBARIUM

Herbarium of The University of Georgia
McINTOSH COUNTY
FLORA OF GEORGIA

Triplasis purpurea (Walt.) Chapm.

FIGURE 2. Herbarium sheet of *Triplasis purpurea* from the University of Georgia Botany Herbarium with diseased specimens.

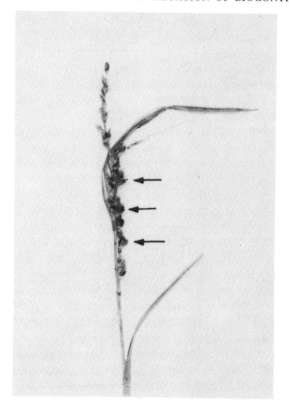

FIGURE 3. Close-up of specimen from herbarium sheet in Fig. 2 showing ovaries infected with smut (arrows).

actual herbarium specimens often can provide useful information since diseased specimens sometimes end up, perhaps unwittingly, on herbarium sheets (Figs. 2 and 3). Herbarium records can also provide clues as to probable location of particular species in the field; this can be especially helpful in an unfamiliar area by saving time that otherwise may be spent in unproductive searching. Information on when to look likewise can be useful. If one wishes to collect mature seeds of a certain weed, a perusal of specimens collected at different dates should reveal the proper time to obtain them, thus saving unnecessary field trips. Locations of herbaria and names of curators can be obtained from the *Index Herbariorum* (15), which is updated periodically.

Thus far these comments have concerned the host, but they apply equally well to the pathogen. Data from mycological herbaria may provide valuable information on the host range of a particular pathogen, and this information may also give clues as to the host specificity of the fungus. Candidates to be used as biocontrols should be host specific, and although strain differences may exist, if the fungus species normally has a wide host range, one may suspect that the candidate fungus also does. Herbarium records may indicate which plant species should be included in host range studies.

Much information on host range and occurrence of particular pathogens is available in State Extension Service Plant Disease Clinic files, but these records are usually of limited availability. Some states such as Georgia publish lists intended mainly for in-house use.

The major difficulty in using data in herbaria is accessibility. In many instances the information is contained in card files or on the specimens themselves. To retrieve these data, one must search these records by hand, a time-consuming and unexciting endeavor. Moreover, the usefulness of making such a search often cannot be predicted. The ideal solution would be to be able to retrieve such data by means of computer, but unfortunately only a few herbaria are presently equipped to provide this service. One of the leaders in this area is Dr. Theodore Crovello, of the University of Notre Dame, who has computerized the data from the specimen labels in the Edward Lee Greene Herbarium (16). A search of these records is a relatively easy process, and an added benefit is that each item of information on the label is separately retrievable. In addition to listing all the specimens in a particular genus or species, one also might wish to know everything that came from a particular area or habitat. This can be accomplished readily.

As far as I am aware, no holdings of any mycological herbarium have been computerized. The National Fungus Collections, USDA Beltsville Agricultural Research Center, Beltsville, Maryland, is in the process of computerizing data from its rust collections, which will make the records much more accessible.

PLANT DISEASE INDEX

The primary source for information on pathogen host range and distribution in the United States is the *Index of Plant Diseases in the United States* published by the USDA as Agriculture Handbook 165 (17). Although issued in its present form in 1960, this publication contains no new entries after 1953. This work, commonly referred to as the *Host Index* or the *Plant Disease Index*, lists all fungi, bacteria, nematodes, and viruses reported to be pathogenic on plants in the United States and U.S. possessions. This reference is the usual starting point for anyone seeking information on plant diseases in the United States and is frequently used as an aid to the identification of diseases. It has two primary disadvantages: it is now 29 years out of date, and entries are indexed only by host. These problems have been overcome partially by the publication of several state host-pathogen indexes in recent years (18–20) along with an index to the genera of pathogens (21) listed in the *Plant Disease Index* (17). Anyone wishing to compile a complete list of species in any particular fungal genus can do so by consulting the works of Saccardo (22,23) and Petrak (24) as well as the *Index of Fungi* (25).

Because it is consulted so widely around the world, thought has been given to updating and computerizing the data in the *Plant Disease Index* (17). One difficulty in such an undertaking is that it is essentially open-ended with regard to entries; namely, there is no accurate way to predict how many entries are in the literature and consequently no way to estimate the amount of work and costs involved, or even whether such a project is feasible in view of the literature explosion in recent years. In 1975 our laboratory was awarded a grant by the USDA to study the feasibility of revising the *Plant Disease Index* (17). To do this, we selected corn (maize; *Zea mays*) as a model crop because it is widely grown, economically important, and is vulnerable to numerous diseases. Various aspects of the study included a comparison of manual versus computer searches of the literature, sources requiring searching, and development of a program for computerizing the data. Seven sources [*Abstracts of Mycology, Bibliography of Agriculture, Journal of Nematology, Nematologia, Phytopathology, Plant Disease Reporter* (now *Plant Disease*), and *Review of Plant Pathology* (formerly *Review of Applied Mycology*)] were hand-searched beginning with the 1950 volumes, and data were recorded on a specially designed data sheet (Fig. 4). Each separate item of data represented by a three-letter code is separately retrievable from the data base. The study showed that updating and computerizing the *Plant Disease Index* is feasible, given sufficient time and funding (26). Manual searching of journals is necessary because new disease reports are not consistently and uniformly indicated by key words, so they cannot be dependably retrieved by computer. One benefit of computerizing the *Plant Disease Index* (17) is that all the data are available for retrieval, not just the primary entries. For example, references, disease names, and geographic location may be retrieved as well as reproductive states and synonyms of pathogens. Any combination of these also may be retrieved. One result of this expanded coverage was to enlarge the four pages devoted to maize in the *Plant Disease Index* (17) to approximately 260 pages printed in three parts in

Source (SOR): Circle. Lit. ref.; Report; Clinic; Herb. spec.; _____

_____

Author (ATI): _____
                       Last   First   Middle

Co-Authors (ATU):

   (1) _____ (4) _____
          First   Middle   Last

   (2) _____ (5) _____

   (3) _____ (6) _____

Article Title (TTL): _____

_____

Journal (PDT): _____

(PDV)_____(PDI)_____(PDP)_____(PDD)_____
         Volume   Issue   Pages   Year

Host Family (HFM): _____

Host: (HGE) _____ (HSP) _____.
                        Genus   Species

        (HAC) _____ (CON) _____
                    Author citation   Common name

Pathogen: (PGE) _____ (PSP) _____
                        Genus   Species

           (PAC) _____ (SYN) _____
                    Author citation   Synonym

Agent Category (CAA): Circle. Fun.; Bac.; Nem.; Vir.; Mycop.; Arth.; Pro.; Alg.;

                        Fl.-plt.; Phys.; Gen.; Non-pat.; Toxin; Unk.

Disease caused (DIS): _____

Location (LOC): _____
                       States and/or territories (USPS abbr.)

_____

Comments (COM): _____
                       (include f. sp.; conidial state; vectors; alternate hosts,etc.)

_____

_____

FIGURE 4.   A data sheet used to collect information for maize disease index.

the revision. Even allowing for the much larger type size and format of the revision, the increase in readily usable information is evident (27–29).

CONCLUSIONS

The cornerstone of any successful research undertaking is ready access to the past and present relevant information concerning that particular project, regardless of whether this information is published in journals or located in files or on herbarium sheets. To the extent that this information is inaccessible, researchers are deprived of data that may be valuable to the success of their research. Unaware of its existence, researchers may even duplicate their work or at least be unable to utilize it. Inaccessible data are of little value. Administrators responsible for funding research need to consider proposals for preparing indexes and bibliographic monographs that make information available to researchers as being equal in value to the research itself. The existence of such sources would actually save valuable time that investigators would otherwise spend updating their references. This is particularly true of research on the biocontrol of weeds with plant pathogens since so much potentially valuable data exist in unpublished form in plant disease clinic files and in herbaria. Research that extracts, compiles, and computerizes these data, making them readily available to everyone, should be encouraged and funded as an integral part of the search for microbial herbicides. Decisions based on outdated references deprive everyone concerned of the maximum benefits modern science has to offer.

I appreciate the assistance of the following University of Georgia staff members in preparing this paper: Arlene E. Luchsinger, Science Bibliographer, for reviewing the manuscript and offering helpful suggestions; and Steven Brown, Reference Librarian, and Edward Warren, Office of Computing Activities, for providing information on computerized abstracting services.

REFERENCES

1. *Biological Abstracts.* 1980. Biosciences Information Service, Philadelphia, PA.
2. *Weed Abstracts.* 1980. Commonwealth Agricultural Bureaux, Slough, England.
3. *A New Journal on Biocontrol.* 1980. Commonwealth Agricultural Bureaux, Slough, England.
4. *This is BIOSIS.* 1978. Biosciences Information Service, Philadelphia, PA.

5. *Automated Retrieval Service for the National Agricultural Library.* 1976. U.S. Department of Agriculture, Beltsville, MD.

6. *BIOSIS Coverage Distribution.* 1978. Biosciences Information Service, Philadelphia, PA.

7. *BIOSIS Training Manual.* 1978. BIOSIS previews edition. Biosciences Information Service, Philadelphia, PA.

8. C. L. Gilreath. 1976. *CAIN Online Users Guide.* National Agricultural Library, U.S. Department of Agriculture, Beltsville, MD.

9. *Guide to the Indexes for 1972 Biological Abstracts and Bioresearch Index.* 1972. Biosciences Information Service, Philadelphia, PA.

10. *CAB News.* Vol. 8. 1980. Commonwealth Agricultural Bureaux, Slough, England, pp. 1–8.

11. *SSIE Science Newsletter.* Vol. 7, No. 7. 1978. Smithsonian Science Information Exchange, Washington, DC, p. 1.

12. R. C. Smith, W. M. Reid, and A. E. Luchsinger. 1980. *Smith's Guide to the Literature of the Life Sciences,* 9th ed. Burgess, Minneapolis, MN, 223 pp.

13. S. B. Jones, Jr. and N. C. Coile. 1979. *List of Georgia Plants in the University of Georgia Herbarium.* Department of Botany, University of Georgia, Athens, GA, 53 pp.

14. K. P. Buchholtz, B. H. Grigsby, O. C. Lee, F. W. Slife, C. J. Willard, and N. J. Volk. 1954. *Weeds of the North Central States.* University of Illinois Agricultural Experiment Station Circular 718, 239 pp.

15. P. A. Holmgren and W. Keuken. 1974. *Index Herbariorum.* Part I, *The Herbaria of the World,* 6th ed. *Regnum Vegetabile* **92,** 1–397.

16. T. J. Crovello. 1972. Computerization of specimen data from the Edward Lee Greene Herbarium (ND-G) at Notre Dame. *Brittonia* **24,** 131–141.

17. Anonymous. 1960. *Index of Plant Diseases in the United States.* Agriculture Handbook No. 165. U.S. Government Printing Office, Washington, DC, 531 pp.

18. L. F. Grand, ed. 1977. *North Carolina Plant Disease Index.* North Carolina Agricultural Experiment Station Technical Bulletin 240, 105 pp.

19. G. K. Parris. 1959. *A Revised Host Index of Mississippi Plant Diseases.* Mississippi State University Botany Department Miscellaneous Publication No. 1. State College, MS, 146 pp.

20. C. Wehlberg, S. A. Alfieri, Jr., K. R. Langdon, and J. W. Kimbrough, 1975. *Index of Plant Diseases in Florida.* Florida Department of Agriculture and Consumer Services, Division of Plant Industry, Gainesville, FL, Bulletin 11, 285 pp.

21. R. T. Hanlin and J. H. Chalkley. 1967. Index to genera of pathogens listed in the *Index of Plant Diseases in the United States. Plant Dis. Rep.* **51,** Part 1, 235–240; Part 2, 323–328; Part 3, 419–424; Part 4, 515–520.

22. P. A. Saccardo. 1882–1931. *Sylloge Fungorum Omnium Hucusque Cognitorum.* Vol. 1–25. Published by the author, Pavia, Italy.

23. P. A. Saccardo. 1972. *Sylloge Fungorum Omnium Hucusque Cognitorum.* Vol. 26 (compiled by E. K. Cash). Johnson Reprint Corporation, New York.

24. F. Petrak. 1930–1944. *Verzeichnis der Neuen Arten, Varietäten, Formen, Namen und Wichtigsten Synonyme.* Parts I–VII. *Just's Bot. Jahresber.* 1952. **48, 49, 56–58, 60, 63.** (Reprinted as *Index of Fungi,* 1920 and 1930. Vols. 1 and 2. Commonwealth Mycological Institute, Kew, England).

25. Anonymous. 1940–present. *Index of Fungi*. Commonwealth Mycological Institute, Kew, England.

26. R. T. Hanlin. 1978. Revision of the *Index of Plant Diseases in the United States*—a feasibility study. *Plant Dis. Rep.* **62**, 377–381.

27. R. T. Hanlin, L. L. Foudin, Y. Berisford, S. U. Glover, J. P. Jones, and L. H. Huang. 1978. *Plant Disease Index for Maize in the United States*. Part I. *Host Index*. Georgia Agricultural Experiment Station Research Report 277, 62 pp.

28. R. T. Hanlin, Y. Berisford, and L. L. Foudin. 1978. *Plant Disease Index for Maize in the United States*. Part II. *Pathogen Taxonomic Index*. Georgia Agricultural Experiment Station Report 285, 101 pp.

29. R. T. Hanlin, L. L. Foudin, Y. Berisford, J. P. Jones, S. U. Glover, and L. H. Huang. 1978. *Plant Disease Index for Maize in the United States*. Part III. *References*. Georgia Agricultural Experiment Station Research Report 294, 104 pp.

14

DISCUSSION OF TOPICS

A. D. WORSHAM

Crop Science Department, North Carolina State University,
Raleigh, North Carolina

This chapter is a discussion of major conclusions from other chapters in this book with a view to presenting a concise critique of the subject of biological weed control with plant pathogens. The organization of this book is very appropriate. The contents reflect the many questions that would usually be asked by scientists and nonscientists concerning this subject. Some of these questions would logically be:

What are the objectives of the book?

Why are new weed management strategies needed, particularly the addition of microbial biological control methods?

Why should this biological control effort be expended on the pest weeds?

What is the current status of weed control with plant pathogens?

Are there native pathogens on weeds that might be useful for biological control?

Are there exotic pathogens that may be useful, and if so, can they be imported?

Could a plant pathogen ever reduce a weed population significantly to be of practical value?

Paper No. 7071 of the Journal Series of the North Carolina Agricultural Research Service, Raleigh, NC 27650.

What are the biological constraints on a control method that is dependent on a disease epidemic for its success?

Are levels of genetic variability in the weed hosts such that success could not be expected with any or all weeds?

Can genetic heterogeneity in plant pathogens be used effectively to achieve weed control without potential hazards to the environment?

Can effective epidemics (epiphytotics) be created in weed populations? If so, what must be done to start them, and what natural systems will be involved?

Can the industrial scientists mass-produce microorganisms for biocontrol?

If biocontrol agents can be mass-produced, can industry profitably do so?

Will governmental regulations be sufficiently favorable to allow and encourage registration of biocontrol agents?

Will biocontrol methods, specifically those relying on plant pathogens, be compatible with current integrated pest management systems?

How can scientists best obtain information about this new biological weed management approach?

What kind of collaboration needs to be established or strengthened between scientific disciplines and private and public agencies to help ensure success of this new venture?

All these questions have been addressed in this book, although not all have precise answers at this time. Enough information is presented, however, to justify the cautious optimism among scientists in this area. The optimism is well founded on sound biological principles; thus we may expect increasing interest and continued progress toward practical utilization of plant pathogens in future weed control technology.

WHY WEEDS?

In Chapter 2 Drs. McWhorter and Chandler leave no doubt as to the value of any effort expended to develop additional weed control methods. In spite of significant advances made in weed control technology, weeds still cause an approximate 12% loss in agricultural production. The losses caused by weeds and the cost of weed control in the United States amount to about $14 billion annually, which is more than losses due to all other pests combined.

Weed scientists have not developed control techniques for many specific weeds. Also, new, major problem weeds that could not be controlled by existing methods have appeared, most often in monocultures after continuous use of specific herbicide programs. Herbicide usage is still increasing, with about 121 million ha treated in 1980, and herbicides constituted 61% of all pesticide sales in the United States in 1978.

Factors further complicating weed management efforts are the many different kinds of weeds with varying periods of germination and highly divergent life cycles. The management of these diverse weed populations requires an integrated systems approach that employs chemical, cultural, mechanical, biological, ecological, and bioenvironmental methods. Unfortunately, research in weed science has not been sufficient to provide management tools to handle such complicated programs in all crops.

Because of their efficiency, herbicides will continue to be a key ingredient in most integrated weed management systems in the foreseeable future. Although few problems have arisen because of this heavy reliance on herbicides, some potential problems exist, and certain newer problems will arise. Some herbicides degrade slowly enough to persist into the next season. This must be taken into account in weed management systems. Contamination of the environment is possible where herbicides are carelessly used. There have been crop injuries from herbicide drift, and herbicides have occasionally been detected in runoff water. But these problems have not been serious. Other existing and potential problems include crop injury through misuse or adverse environmental conditions, human exposure during application, disposal of containers, and increase in weed resistance. Biocontrols, especially with plant pathogens, would offer some solution to some of these problems. Biocontrols will not, however, solve all weed problems.

There are many future research needs for the development of totally effective weed management systems. These research areas and management strategies that we must explore include cultivar interactions with weeds and/or herbicides, levels of weed infestation and thresholds for economical control steps, rate of weed spread, weed-crop competition, allelopathics, physiology of weed seed germination, controlled release herbicides, better application equipment, weed biology, and the projected effect of changing tillage practices on weeds and management practices.

The use of geese for weeding in certain crops and goats for brush control in firebreaks has been attempted in the past and revived from time to time but still appears to have only limited usefulness. The use of fish for control of aquatic weeds, however, is already an example of a very useful biological weed control method (1). However, some of these agents may have only limited applications in biocontrol of weeds. The use of geese in cotton fields in the Mississippi Delta states, for example, was limited by the amount of labor required to manage them and the incompatibility of geese with certain other pest management practices such as insecticide use (2).

Dr. Leroy Holm (3) led an interesting discussion at this workshop (see the Preface to this book) on weed terminology as it related to the importance of different weed species worldwide and how we would determine which ones to choose first as biological control targets:

Nature places particular plant species in our local fields, in arable agriculture over large regions, and sometimes over the crops of the world, by rules which are not yet clear to us. Wherever they are, we describe them in loose terms such as problem weeds, troublesome, major, important, etc., but these words are never limited or defined. Biological control work is very tedious, very expensive, and it takes a long time. Which class of weeds ought we to work on *first?* Are *troublesome* weeds the same as *important* weeds? In Indonesia prior to the last period of internal strife, *Imperata cylindrica*, the worst perennial grass of southeast Asia, was well under control on the major rubber plantations. This was because there were separate company divisions just for the control of this weed. It was not *troublesome!* But it waited at the edge of each plantation. During the difficult political upheaval which followed, such farms were neglected and there were few funds for weed control. *Imperata* very quickly invaded and took over between trees. It always was the most *important* rubber weed of southeast Asia, and it had now become *troublesome*, again, in Indonesia. It took many years to restore the plantations.

Quackgrass (*Agropyron repens*) waits in every fencerow of the northern corn belt in the United States. It is the worst temperate perennial grass of the area. Just now, with the use of herbicides, machines, and rotations the corn fields can be quite free of it—it is *not troublesome*. But it is very *important!* If the herbicides are withdrawn, if energy sources for proper tillage become too costly, etc., quackgrass, the most *important* weed, will return and it will be *troublesome*.

STATUS OF WEED CONTROL WITH PLANT PATHOGENS

Dr. Templeton in Chapter 3 has indicated that at the end of the initial decade of modest research and development efforts with plant pathogens for weed control, there are projects in 13 countries and 18 states in the United States. Approximately 83 pathogens are being studied to control 54 target weed species. Outstanding successes have been achieved in the control of skeletonweed (*Chondrilla juncea*) in Australia and hamakua pamakani (*Eupatorium riparium*) in Hawaii. The use of plant pathogens as microbial herbicides (see Chapter 3 for definition of microbial herbicide) is approaching commercialization for two target weeds: water-hyacinth (*Eichhornia crassipes*), and northern jointvetch (*Aeschynomene virginica*); it has reached commercialization in the case of strangler vine (*Morrenia odorata*).

INFLUENCE OF DISEASES ON PLANT POPULATIONS

The impact diseases can have on plant populations has been discussed by Dr. Quimby, who has cited some serious epiphytotics that have caused severe economic and social consequences around the world (Chapter 4). However, reference books on ecology rarely address plant diseases as important determinants of the constituency of plant communities. Plant pathogens have rarely been studied as part of an ecological system in nature, and their relevance to population biology and population genetics is mainly an unexplored territory.

Potato late blight in Ireland, coffee rust in Ceylon, and chestnut blight and oak wilt in the United States have been listed as examples of uncontrolled epiphytotics. Factors that apparently favored the occurrence of several major epiphytotics and the theories of coevolution that relate to these events were discussed in Chapter 4. In this chapter Dr. Quimby has left the thought that the potential for manipulation of pathogens to control weeds is very promising.

When considering these epiphytotics in terms of relating underlying principles to biocontrol of weeds, the question arises as to the effectiveness of searching for exotic pathogens to control a native weed versus searching for native pathogens. Coevolution of weeds and their pathogens has resulted in endemism (4). Endemism would result in native pathogens being innocuous against native weeds, but many opportunities exist for human intervention in favor of the pathogen. However, despite coevolution resulting in endemism of pathogens, the potential for discovering new diseases on native weeds cannot be ruled out. In a period of 12 months, Walker (5) discovered two new diseases on weeds in Mississippi.

Some of the more devastating epiphytotics in natural plant populations have occurred on members of the climax communities, namely, the tree species (6). The idea that most annual plants may generally escape diseases by their mobility—through their dispersal patterns—has been suggested from surveys of plant communities at different successional stages (7). Generally, more diseases have been found in the climax stages of succession, and few diseases are found on weeds. For example, in a survey in Texas of various plant communities of different ages, it was found generally that the older the succession, the more stable was the community and more diseases were found (7). It is possible that annual plants escape disease in time and space. The annuals "move around" too quickly for diseases to "catch" them. Although such an idea is reasonable in terms of insect attack, it may not be so reasonable in terms of pathogens that are everywhere.

CONSTRAINTS ON DISEASE DEVELOPMENT

Chapter 5 is a discussion of the basic plant pathological principles used in this area of work. Dr. Holcomb has addressed some of the constraints that would affect weed control with plant pathogens. These constraints include unfavorable temperature, resistance factors, spatial continuity of the hosts, host geographic range and distribution, diversity of host genetic base, and age and vigor of host plants. The goal of plant pathology is to understand the disease complex so that plant diseases can be managed or controlled. Plant pathological principles can also be used to give advantage to the pathogen in the case of biocontrol of weeds with plant pathogens. By understanding the constraints on plant pathogens and removing or manipulating those constraints, one might achieve more success in biological weed control.

One of the major ways that diseases are constrained in nature is through host resistance. However, information on the aspect of host resistance in natural communities is lacking. It is not known how long it takes plants to develop resistance or what the rate of genetic change is in natural plant populations. All the natural communities have evolved some resistance to all of the diseases that are in their environment, but the underlying genetic mechanisms are unknown. After 25 years of intensive herbicide use, there are a few examples of triazine-resistant subspecies. Although the factors underlying inheritance of herbicide resistance are known, it is not known whether this resistance developed genetically or if the few resistant individuals in a population have only now increased to noticeable proportions (7). One reason why we do not know more about the genetics of natural populations is because scientists have been very reluctant to work on plants that are not crops, especially weeds. The levels of genetic variation and resistance in weeds may influence the length of time necessary for exotic weeds to succumb to a native pathogen, assuming the two had not encountered each other before. Saltcedar (*Tamarix pentandra*) in Argentina might serve as an example (see Chapter 4). Saltcedar is a Mediterranean plant taken to South America where a 1–2-year-old planting was wiped out by *Botryosphaeria tamaricis*. The host probably consisted of a very homogeneous genotype, but we do not know whether the pathogen was transported with saltcedar or was native to South America. In any case, it did not take very long for the pathogen to decimate the host.

GENETIC VARIATION IN WEEDS

Since genetic diversity of the weed host would be a constraint to biocontrol with plant pathogens, Dr. Barrett has discussed the genetic variation and factors influencing genetic variation in weeds in Chapter 6. The method of reproduction of the weed is important to success in biological control. Limited evidence from surveys involving biocontrol agents suggests that asexually reproducing weeds are more easily controlled than sexual species. This difference may be associated with the lower levels of genetic variation present in weed species with restricted recombination systems. Generally those weeds that are inbreeders can be controlled successfully with biological agents, whereas weeds with sexual outbreeding systems are difficult to control biologically. This simply relates to the amount of genetic variation generated by the method of reproduction. *Opuntia* spp. are apomictic plants that have virtually no variation within the populations in Australia; they have a very limited genetic base. Therefore, it is not surprising that the Australians obtained good biological control of cacti with *Cactoblastis cactorum*. A similar situation exists with skeletonweed in Australia, where successful control has been achieved with the rust *Puccinia chondrillina*.

The best targets for biocontrol appear to be apomicts such as skeletonweed, pamakani, St. Johnswort (*Hypericum perforatum*), cacti (*Opuntia* spp.), and Hawaiian blackberry (*Rubus penetrans*). Aquatic weeds that reproduce predominantly by clonal propagation such as alligatorweed (*Alternanthera philoxeroides*), elodea (*Elodea canadensis*), waterhyacinth, hydrilla (*Hydrilla verticillata*), waterlettuce (*Pistia stratiotes*), and giant salvinia (*Salvinia molesta*) would also seem to be excellent targets for attempts at biocontrol.

It is quite possible that in many weed species the actual amount of genetic variation in populations will be a poor predictor of whether successful control can be achieved. What is probably more important is the nature of the genetic variation present in the population and how this will influence the rate at which host resistance is likely to evolve under field conditions. Unfortunately, at this time little is known of the evolution and maintenance of host resistance in natural plant populations.

The question of potential for development of resistance in weeds to biocontrol agents cannot be answered at this time. However, it is en-

couraging to note that despite extensive and repeated use of some herbicides over the past 30 years, there are relatively few reports of evolution
of genetic resistance to herbicides.

ADVANTAGES AND HAZARDS OF GENETIC
VARIATION IN PATHOGENS

In Chapter 7 Dr. Leonard has discussed some potentials for hazards as
well as benefits from using plant pathogens as weed control agents. First,
in the case of the classical biocontrol tactic (see Chapter 3), it is probably
safer to screen potential host plants against a heterogeneous population of
a pathogen than against a single isolate chosen for importation. A mixture
of pathogen races or strains used in screening is more likely to reveal the
potential hazards of genetic heterogeneity in plant pathogens used for
biocontrol. Second, when a facultative parasite is considered for introduction into a new area of the world, it is important that the pathogen be
avirulent on useful plants, not just weakly virulent. Adaptation to a new
host might occur as it does in the case of *Phytophthora infestans* from
potato, which becomes highly virulent to tomato after five or six passages
through tomato leaves. For this reason, it would seem prudent to avoid
importing new pathogens for weeds that are closely related to valuable
plant species. There is a real danger that the pathogen may be preadapted, so that with only a single genetic change it might attack a new,
nontarget host.

Most of the potential hazards of exotic pathogens can be avoided by
selecting pathogens that are already endemic in the area where they will
be used. Genetic variability in an endemic pathogen used as a bioherbicide should pose little or no hazard, especially if the bioherbicide is used
against annual plants in a temperate climate and the pathogen has limited
ability to survive in the soil.

Genetic heterogeneity can be a useful feature in plant pathogens chosen as biocontrol agents. In the classical approach, it should be possible to
find an isolate with the desired virulence if the pathogen populations were
genetically heterogeneous. In the bioherbicide approach, the pathogens
could be bred and adapted for virulence and other specialized characteristics, much like the industrially useful microorganisms. Mutants could be
induced to increase virulence. The existence of quantitative genetic variation for virulence in pathogen populations permits the selection of isolates
with greater virulence through recombination of existing genes.

In the unlikely event that weeds do become resistant to plant pathogens used as biocontrol agents through natural selection, the pathogen can be selected periodically on the weed population to increase its virulence to match the increased resistance in the host.

CREATING EPIDEMICS

Dr. Shrum has discussed the dynamics of epidemics in Chapter 8. For epidemics to occur, in general, maximum reproduction of the pathogen should be favored by disseminating the inoculum early in the growing season. This permits the maximum number of infection cycles of the pathogen and takes advantage of the generally moist and moderate weather of spring. An effective amount of inoculum should be applied initially. The amount of inoculum is multiplied through each new infection cycle. The inoculum must be applied when the infection window is open to permit maximum infection. The infection window includes both compatibility of climate and susceptibility of host. Care should be taken in biocontrol efforts not to culture and disseminate hyperparasites inadvertently along with the desirable biological agents.

Further research is needed to understand how best to manipulate epidemics to achieve weed control. More information is needed to determine how soon the full epidemic must occur and what level of infection and rate of spread of the epidemic will result in the best degree of weed control.

MASS PRODUCTION OF MICROORGANISMS FOR BIOLOGICAL CONTROL

In Chapter 9 Dr. Churchill has described experiments in mass production of microorganisms for biological control of weeds. Experience over the last 10 years has shown that large quantities of viable, dried spores of *Colletotrichum gloeosporioides* f. sp. *aeschynomene* (CGA) and *C. gloeosporioides* f. sp. *jussiaeae* (CGJ) can be produced that will give good control of the respective host weeds. The CGA and CGJ spores produced under the fermentation conditions were not the type produced naturally on the host weeds, but these spores were pathogenic. The author suggests that industries could supply the required amount of viable spores at a price that would appear to be competitive with that of the chemical pesticides now being used. However, since the mass production of microorganisms for biocontrol of weeds is highly unpredictable, plant pathologists, microbiologists, and engineers will have to cooperate fully to solve the many problems that may arise.

Dr. Churchill has discussed all aspects of mass production: stock culture preparation; selection of culture medium; standardized inoculum; shake flask fermentation; methods of maintaining sterility; scaling up the fermentation from shake flask to production tanks; innovations in the modern fermentation production plant; spore production by fungal pathogens; harvesting and drying spores; and germination of dried spores.

Mass production of microorganisms is a very complicated process. Critical factors must be determined for each strain such as the carbon source, nitrogen source, carbon:nitrogen ratio, mineral supplements, pH, temperature, and agitation and aeration. For example, CGA and CGJ belong to the same genus and species but differ in agitation and aeration requirements.

Modern fermentation plants have automatic weighing and dispensing of bulk ingredients directly into fermentation tanks. Continuous automatic measurements of pH, airflow rate, agitation speed, oxygen uptake, dissolved oxygen, and power input are also common. The end result is vastly more efficient fermentation and reduced production costs. These improvements should give assurance that the versatility and economical operation are available for the mass production of spores of bioherbicides at a cost that is competitive with those of chemical herbicides.

COMMERCIALIZATION OF MICROBIAL BIOLOGICAL CONTROL AGENTS

Chapter 10, by Dr. Bowers, begins with a somewhat pessimistic review of previous efforts to commercialize biological control agents. Dr. Bowers has noted that only ten biological control agents are currently registered by the Environmental Protection Agency (EPA) as compared with 1500 registered chemical pesticides. Also, the historical achievement of the worldwide rate of "complete success" in classical biological control of arthropods has not been an incentive for industry to commercialize either macrobial or microbial agents.

However, several studies cautiously project modest gains in the use of biologicals, provided that several factors continue to impinge on pest control methods. Some factors that favor the use of biological control agents are enhancement of integrated pest management systems, continued government support of research and development of biological controls, continued enactment of stringent regulations of chemical pesticides, and adoption of new regulations governing the registration of biocontrol agents.

Bowers has discussed the four phases of commercialization of biological control agents—discovery, lead development, product development, and marketing. The nonindustrial agencies will have to shoulder most of the responsibility for discovery of biological control agents. The importance of patents has been emphasized, and nonindustrial biological scientists are advised of the importance of protecting discoveries with patents. The interface between nonindustrial and industrial scientists will be an essential key to accelerating progress in the use of biological control agents for pest management.

Once an effective organism is discovered, many factors must be studied to determine whether it can be commercialized. These factors include efficacy, safety, specificity, genetic stability, potential for mass production, formulation, product stability, and shelf life.

The product must be profitable, so economic analyses must be conducted before significant funds are invested. These analyses include the market potential, research and development costs, capital investments, finished-goods cost, royalties, and return on investment (ROI). Product development involves gathering of data on a use pattern, safety parameters, and assurance of a high-quality final product. These will all be more difficult for a biological control agent than a chemical agent.

It is reasonable to expect that the free enterprise system will develop biocontrols rapidly in the coming years. However, the decision to commercialize biocontrol agents will be made by market analysts, accountants, and managers, not by scientists. Therefore, if a product will not return anything on the investment within 13 years, industry will not attempt to develop it (8). With classical biocontrol agents, it might be the governmental agencies' role to fill the needs where it is not profitable for industry to develop or distribute these agents. Some examples of the latter category have been the work on rush skeletonweed, musk thistle (*Carduus nutans*), and yellow starthistle (*Centaurea solstitialis*) (9). The Animal and Plant Health Inspection Service (APHIS) of the U.S. Department of Agriculture (USDA) has demonstration programs on biological control of certain insects and weeds (9). The general opinion of the industry representatives appears to be that the microbial herbicide approach offers the greatest potential for industry involvement. Bowers has summarized by saying that industry will commercialize microbial pest control agents, but only those ventures that show promise of being profitable will be actively pursued.

REGULATION

In Chapter 11 Dr. Charudattan has outlined some proposed and current regulations, mainly in the United States, concerning the use of plant pathogens as weed control agents. Regulations are necessary and must be sufficiently stringent to protect the public, but not so excessive as to stifle and delay progress in this field. Regulations and guidelines should also be sufficiently flexible to allow for biological differences among plant pathogens and the pathogen-host systems involved in biocontrol.

Importation and release of exotic plant pathogens in the United States for biological control purposes are regulated under the Plant Quarantine Act of 1912 and the Federal Pest Control Act of 1957. The Plant Protection and Quarantine (PPQ) section of the USDA's APHIS has the responsibility to issue permits for importation of exotic plant pathogens. An advisory group of scientists, called the Working Group on Biological Control of Weeds, established by the USDA and the U.S. Department of Interior, reviews proposals and provides recommendations to the researchers and the PPQ on the quarantine testing and release of exotic biological agents to control weeds.

The use of plant pathogens as weed control agents, either as classical biocontrol agents or as microbial herbicides, will be regulated and registered by the EPA under the Federal Insecticide, Fungicide, and Rodenticide Act (FIFRA), as amended. In 1980 the EPA proposed guidelines for registering biological pesticides (given the term "biorational pesticides"), which pending review, will become law. These guidelines specify the data required and give instructions to the registration applicants and the public concerning registration of microbial pest control agents. The guidelines also discuss current registration procedures, data and labeling requirements for registration, standards for product efficacy, acceptable test methods for the development of required data, and information required in test reports.

In developing the guidelines, the EPA has taken a positive approach toward biological control agents, and this should aid in the orderly availability of microbial plant pathogens for weed control.

INTEGRATION OF MICROBIAL BIOCONTROL AGENTS WITH EXISTING PEST MANAGEMENT PROGRAMS

In Chapter 12 Dr. Smith has pointed out that although plant pathogens can be used to control weeds, considerable research will be required before microbial biocontrol agents can be used judiciously in integrated weed management systems.

Integration of the bioherbicide approach to weed control with weed management systems for rice requires that the weed pathogens be compatible with the chemical herbicides, fungicides, and insecticides in use. Mixtures of weed pathogenic fungi and many of these pesticides have been shown to be incompatible and could not be employed together. However, by precise timing of sequential applications, good results on target pests have been obtained. On the other hand, tank mixtures of two or three weed pathogenic fungi have been used successfully in Arkansas.

Dr. Smith has given an example of how weed pathogenic fungi could be incorporated into a pest management system for a rice-soybean cropping system. Crop management is also an important component of the system to be considered. For example, for CGA to be effective, the soil must be saturated with water when the spores are applied. Therefore, applications of CGA should be timed to coincide with rain or irrigation for effective results.

Possibilities for incorporating three weed pathogenic fungi into a soybean-cotton cropping system have also been discussed. For example, during the soybean production year, CGA and *Colletotrichum malvarum* (CM) may be applied in tank mixtures or sequentially for control of northern jointvetch and prickly sida (*Sida spinosa*). Likewise, in the cotton production year, CM and *Alternaria macrospora* (AM) may be applied for control of prickly sida and spurred anoda (*Anoda cristata*). The three fungi may be used alone or in combination for control of these three weeds in these two crops.

INFORMATION SOURCES FOR THE BIOCONTROL RESEARCHER

The problems of searching the literature for pertinent references on the subject of biological control of weeds are compounded because two groups of organisms are involved: the host weed and the pathogen. Sources of information may be quite different for each organism, and these have been discussed in Chapter 13 by Dr. Hanlin. The primary abstracting journals covering these areas are *Abstracts of Mycology, Review of Plant Pathology*, and *Weed Abstracts*.

For more complete and current information, computer literature retrieval searches are available. The two most widely used computerized abstracting services applicable to agriculture are the BioSciences Information Service (BIOSIS) and the Agricultural On-Line Access System (AGRICOLA). The Commonwealth Agricultural Bureaux (U.K.) also provides a computerized version of citations from the 26 abstracting journals that it searches. Workers in federal or land grant institutions also have access to the USDA Current Research Information Services and the Current Awareness Literature Service data bases. Two additional services that may be purchased are the Smithsonian Science Information Exchange (SSIE) and the SSIE Notice of Research Project.

Dr. Hanlin has also advised researchers in the biocontrol area not to overlook the information contained in sources such as herbarium records and host indexes. Much useful information can be obtained on both weed and pathogen. The major difficulty, however, in using the data in either herbarium or host indexes is the lack of accessibility to many researchers. Administrators responsible for funding research are urged to support proposals for the preparation of indexes and bibliographic monographs that make this information available to researchers. These are of equal value as the biocontrol research itself.

ESTABLISHING AND STRENGTHENING COOPERATION

Biological control of weeds with plant pathogens is a new and emerging technology. Dr. Warren C. Shaw pointed out at the workshop that cooperation between disciplines and organizations involved with biological control will be essential to the success of this new technology (10). Dr. Shaw reviewed some of the problems that agriculture will face in the future: "Until the year 1830, the world's human population was only about one billion. But it is estimated that by the year 2000, there will be seven billion people in the world. Just 30 years from now the world's population will double. This means that within the next 30 years agriculturalists must produce as much food as they have produced since the beginning of man's history." One may question whether this many people can be fed. Dr. Shaw believes they can, but expanded research to develop improved technology will be required to meet the challenge. The use of plant pathogens for the selective control of weeds will help meet the challenge (10).

Dr. Shaw reviewed the magnitude of the weed problem in the United States and emphasized that "despite the use of the best weed management technology currently available, agricultural losses caused by weeds are estimated to be 10 to 12% or more than $10 billion annually. Annual cost of weed control is about $6.2 billion. Thus the losses and cost of control are estimated at more than $16 billion each year."

He also pointed out that an important mission of agricultural research is to develop the technology needed to reduce the losses and costs of weed control and assure an adequate supply of nutritious food, high-quality feed and fiber, and a quality environment. He feels that "No mission is of greater importance to the general public, and this mission could not be accomplished unless agricultural yields are increased and crop production be made more efficient. However, agricultural yields have begun to level off globally. Even the most elementary analysis will show that an increase in agricultural yields cannot be achieved without effective integrated pest management systems (IPMS) that utilize the best combination of principles, practices, technologies, and strategies" (10).

The use of plant pathogens for selective weed control can help to reduce the losses caused by weeds and the cost of weed control, increase the yields and quality of crop, and reduce the need for tillage—thus reducing soil erosion and energy requirements. The use of plant pathogens for weed control, where successful, will also strengthen integrated weed management systems and IPMS (10).

Dr. Shaw mentioned that there are probably less than 20 scientists in state, federal, and industrial organizations in the United States and perhaps just as many in other countries devoting their efforts to research on the development of plant pathogens for selective weed control. Therefore, these few are truly pioneers. In the United States, Dr. H. L. Walker at the Southern Weed Science Laboratory, Stoneville, Mississippi is the first USDA scientist ever employed full-time to develop plant pathogens for selective weed control. Although only a few scientists are engaged in research to develop plant pathogens for weed control in the United States, there are more than 200 weed scientists and 800 plant pathologists with related talents that could be used in part to support this type of research (10). Dr. Shaw believes that there is a great need for newer technology to control weeds that are not controlled by current methods. Also, a diversity of control technology must be used to prevent the development of tolerance in weeds to currently used herbicides and to stop undesirable ecological shifts from controllable weed species to those that are uncontrollable by current technology.

In spite of the dearth of scientists working on plant pathogens for selective weed control, successes in this area have been truly remarkable in view of the amount of scientific time and funds devoted to this research. Dr. Shaw pointed out that only some 15 scientists were involved in the development of plant pathogens for control of strangler vine in citrus groves, northern jointvetch in rice fields, spurred anoda in cotton and soybean, rush skeletonweed in grazing lands and grain fields, and aquatic weeds.

These examples have clearly demonstrated that plant pathogens can be developed for selective weed control. As microbial herbicides, pathogens can be formulated and stored. They can also be applied with ground or aerial equipment as foliar or soil treatments. Finally, it has been demonstrated that pathogens will control weeds effectively, economically, and safely without causing crop damage or environmental problems (10).

The development of weed science has always been characterized by strong multidisciplinary and multiinstitutional cooperation (10). Currently it is fashionable to refer to the special partnerships that exist between the USDA and the states in agricultural research. However, for weed scientists, this concept of cooperation falls short of reaching the desired objective. Progress in developing weed management systems has been dependent on cooperation among federal, state, and industrial scientists in the disciplines of agronomy, aquatic biology, botany, chemistry, ecology, entomology, engineering, horticulture, plant physiology, plant pathology, toxicology, wildlife biology, and others.

The current success in developing plant pathogens for weed control has been uniquely characterized by strong multidisciplinary and multiinstitutional cooperation. Future success will depend on continued and improved cooperation.

Dr. Shaw (10) felt that the future progress in this area would be determined largely by the resources allocated to basic and applied research and the success in stimulating multidisciplinary and multiinstitutional cooperation for technology development and use. He indicated optimism for success. Some of the steps that Dr. Shaw proposes to ensure success are:

1. Organizing weed science–plant pathology committees in the Weed Science Society of America and the American Phytopathological Society and perhaps carrying such a group into a formal organization.

2. Organizing more regional research projects (see the Preface to this book) or individual research projects devoted to biological control of weeds with plant pathogens.

3. Organizing weed science–plant pathology research projects in the state agricultural experiment stations.

4. Achieving a better balance of scientists from different disciplines on teams involved in foreign exploration by assigning weed scientists and plant pathologists to such teams.

5. Broadening the objectives of biological control research funding in all institutions to include development of pathogens for weed control.

6. Establishing a USDA–state–industry–U.S. EPA task force to assure progress in the registration of pathogens for weed control.

7. Devising procedures for assuring the involvement and participation

of industrial scientists at an earlier stage of the planning and conduct of research on biocontrols.

8. Sponsoring a conference to delineate and describe the research roles of USDA, state, and industrial scientists.

9. Organizing a research planning conference (similar to the one that initiated this book; see Preface) of federal, state, and industrial scientists to establish national research priorities so that these needs may be reflected in federal, state, and industrial research budget requests.

10. Developing within the USDA a patent policy that would allow for the issuance of exclusive licenses on public patents.

11. More resources allocated to all areas of basic and applied research on biocontrols (10).

POTENTIAL RESEARCHABLE AREAS

Dr. John F. Fulkerson pointed out at this conference the potential researchable areas that need to be explored for increasing knowledge and improving the prospects for implementation of biocontrols (11). Some of these areas are (1) searching for chemicals as aids to biocontrol measures, (2) studying the possibility of gene substitution in weeds to enhance biocontrol, (3) exploring ways for more incentives to industries to become interested in microbial biocontrol agents, (4) making use of recombinant DNA techniques for selection of more desirable microbial biocontrol agents, (5) initiating studies of subtle balance systems already in operation in nature that tend to keep weeds in check, and (6) developing methods for biological containment of weeds (11).

Dr. Leroy Holm (3), drawing on his many years of wisdom and experience in working with weeds the world over, left the following sobering thoughts with the participants of the workshop:

The promise of biological weed control is exciting and we are encouraged, but realistically our resources, our lack of manpower, and our potential for solving very many specific weed problems in the near term is sobering. We must choose wisely. Should we choose the interesting ones, those we chance upon, or should we safely seek to work on peripheral improvements for projects which seem to be popular? Perhaps what this discipline needs is to strike out boldly for the biological control of *Eichhornia crassipes* (waterhyacinth), *Cyperus rotundus* (nutgrass), the world's No. 1 weed, *Imperata cylindrica, Agropyron repens, Echinochloa crus-galli,* the worst weed of rice, the world's largest crop, *Hydrilla verticillata, Sorghum halepense,* and a host of other *important* weeds. Through these efforts more children will have food, costs will be reduced, and the environment will be spared large amounts of chemicals.

REFERENCES

1. B. Stott and B. R. Buckley. 1978. Costs for controlling aquatic weeds with grass carp as compared with conventional methods. Pp. 253–260 in: Anonymous, ed., *Proceedings of the European Weed Research Society 5th Symposium on Aquatic Weeds*. P.O. Box 14, Wageningen, The Netherlands.

2. C. G. McWhorter. 1980. Southern Weed Science Laboratory, U.S. Department of Agriculture, Science and Education Administration, Agricultural Research, P.O. Box 225, Stoneville, MS 38776, personal communication.

3. L. Holm. 1980. Plant Physiologist, 714 Miami Pass, Madison, WI 53711, personal communication.

4. J. R. Harlan. 1976. Diseases as a factor in plant evolution. *Annu. Rev. Phytopathol.* 14, 31–51.

5. H. L. Walker. 1980. Research Plant Pathologist, U.S. Department of Agriculture, Science and Education Administration, Agricultural Research, Southern Weed Science Laboratory, P.O. Box 225, Stoneville, MS 38776, personal communication.

6. J. G. Horsfall and E. B. Cowling. 1978. Some epidemics man has known. Pp. 17–32 in: J. G. Horsfall and E. B. Cowling, eds., *Plant Disease: An Advanced Treatise*. Vol. 2. *How Disease Develops in Populations*. Academic Press, New York.

7. S. C. H. Barrett. 1980. Professor of Botany, University of Toronto, Toronto, Ontario, Canada M5S 1A1, personal communication.

8. D. S. Kenney. 1980. Section Head, Microbial Products Research, Abbott Research Center, Oakwood Road, Box 173, Long Grove, IL 60047, personal communication.

9. Anonymous. 1980. *Guidelines for PPQ Action Programs in Biological Control.* U.S. Department of Agriculture, Animal and Plant Health Inspection Service, Plant Protection and Quarantine. U.S. Government Printing Office. 0-310-945/APHIS-230, Washington, DC, 19 pp.

10. W. C. Shaw. 1980. National Research Program Leader, Weed Science and Agricultural Chemicals Technology, U.S. Department of Agriculture, Science and Education Administration, Agricultural Research, National Program Staff, Beltsville, MD 20705, personal communication.

11. J. F. Fulkerson. 1980. Plant Pathologist, U.S. Department of Agriculture, Science and Education Administration, Cooperative Research, Room 6444-South, Washington, DC 20250, personal communication.

APPENDIX 1

ABSTRACTS OF
RESEARCH PROJECTS

ISOLATION OF FUNGI PATHOGENIC TO WEEDS

R. T. HANLIN

Department of Plant Pathology, University of Georgia, Athens, Georgia

Before naturally occurring fungal pathogens of weeds can be tested for their use as potential mycoherbicides, they must be isolated and grown in pure culture. Procedures for isolating nonobligate fungal parasites from weeds are illustrated, using a leaf spot of johnsongrass (*Sorghum halepense*) as an example. Infected lesions of johnsongrass are cut into small pieces (≤ 1 cm^2) and are surface-disinfested by submersion in a bleach:ethanol:water (10:10:80) solution for 2 minutes. Leaf pieces must not be left in the solution too long, or internal fungi will also be killed. Surface-disinfested pieces are plated on a suitable agar medium such as Martin's Rose Bengal-Streptomycin medium and are allowed to incubate at room temperature. After 7–10 days a variety of fungi will have grown out from the leaf pieces. Transfers of each different colony are made onto agar plates containing suitable medium such as malt extract, V-8 juice, or potato dextrose. Once sporulating pure cultures of the different isolates are obtained, they can be identified, and those considered likely candidates for use as mycoherbicides can be studied further.

IDENTIFICATION OF PLANT PATHOGENIC FUNGI

R. T. HANLIN

Department of Plant Pathology, University of Georgia, Athens, Georgia

A wide variety of fungi including pathogenic and nonpathogenic species normally will be isolated from diseased tissues when they are plated on agar media. Since only pathogens are useful as mycoherbicides, isolates obtained must be identified at least to genus, and potential pathogens must be selected for testing. Six of the

237

isolates from johnsongrass (*Sorghum halepense*) are illustrated, and the primary characters used in identifying them are pointed out. The six genera are *Fusarium*, *Exserohilum*, *Bipolaris*, *Glomerella*, *Aspergillus*, and *Curvularia*. Two of these genera, *Bipolaris* and *Exserohilum*, are known to have species that cause leaf spot diseases, and they are likely suspects as the cause of the leaf spot on johnsongrass. To test these species, they must be cultured and sufficient spores obtained to inoculate host plants and carry out Koch's postulates, that is, to inoculate host plants in the greenhouse, obtain the disease symptoms originally observed in the field, and then recover the same fungus from the diseased tissue. Once this has been done, the potential of the fungus as a mycoherbicide can be determined.

BIOCONTROL OF WEEDS WITH FUNGAL PLANT PATHOGENS: FOUR DECISIONS MARK THE RESEARCH PROGRESSION

H. W. SPURR, JR. AND C. G. VAN DYKE

Tobacco Research Laboratory, U.S. Department of Agriculture, Science and Education Administration, Agricultural Research, Oxford, North Carolina

Departments of Botany and Plant Pathology, North Carolina State University, Raleigh, North Carolina

The following four decisions briefly illustrate primary steps in the research progression for biological control of weeds with fungal pathogens:

Decision 1—select a target weed. This results after determining whether the weed is introduced or native, whether control is currently available and is economically and environmentally desirable, and whether biological control is an attractive possibility.

Decision 2—select a fungal pathogen. Base this on information obtained from plant disease clinics, plant disease indexes, discussions with plant scientists, and field surveys. Both endemic and exotic fungal pathogens should be considered. Desirable characteristics of a fungal pathogen for weed control include extreme virulence; environmental adaptability; and spores that germinate, disseminate, and overwinter extensively.

Decision 3—select a strategy. There are two main strategies to consider. The classical strategy, often referred to as the natural or self-sustaining strategy, is best suited for introduced weeds in low-value areas such as pastures or waterways where partial control may be sufficient. Obligate fungal parasites are desirable for this strategy because they often become epidemic following a single or limited introduction. Usually a dry formulation of spores is introduced by ground or aerial application in programs coordinated largely by public agencies. The bioherbicide strategy or artificial strategy is usually suited to high-value crops or areas and requires annual application, and complete demise of the weed is desired. A facultative saprophytic fungus that produces spores in fermentative or liquid cultures and can be formulated for spray application is desirable. Cooperation between public and private sectors is required for this strategy to succeed. Both strategies require extensive host range studies.

   Decision 4—continue or terminate the biological control project. How successful or how efficient is the control? Does growth retardation or demise of the weed under field conditions remain economically attractive? Will this control integrate with pest management programs? Currently successful biological control programs have progressed by means of these decisions and illustrate that prevalent skills and resources are sufficient to pursue weed control research with fungal pathogens.

EVALUATION OF *ALTERNARIA ALTERNANTHERAE* AS A CONTROL FOR ALLIGATORWEED

G. E. HOLCOMB

Department of Plant Pathology and Crop Physiology, Agricultural Experiment Station, Louisiana State University, Baton Rouge, Louisiana

A previously undescribed leaf spot disease of alligatorweed (*Alternanthera philoxeroides*) was observed in Louisiana in 1975. Leaf spots were purple bordered with tan, necrotic centers and ranged up to 4 mm in diameter. Elongate stem spots were also present but were fewer in number. A fungus belonging to the genus *Alternaria* was consistently isolated from leaf spots and proved pathogenic in all tests. The fungus was described as a new species, named *Alternaria alternantherae*, and was evaluated as a potential biological control for alligatorweed. Spore suspensions or spores plus mycelium caused typical leaf spots about 3 days after spray application. Defoliation (50–80%) resulted 1 week after spraying, except for the terminal leaf whorl, which was resistant to infection. Damage to stems was negligible. By 4–6 weeks after inoculations, new growth of alligatorweed had masked the effects of fungal defoliation. Leaf spot numbers returned to levels that had existed prior to the artifical inoculations. It was concluded that this fungus was probably a poor candidate for a biocontrol agent.

BIOLOGICAL CONTROL OF WATERHYACINTH WITH THE FUNGAL PATHOGEN *CERCOSPORA RODMANII*

T. E. FREEMAN, R. CHARUDATTAN, R. E. CULLEN, AND D. S. KENNEY

Plant Pathology Department, University of Florida, Gainesville, Florida
Chemical and Agricultural Products Division, Abbott Laboratories, North Chicago, Illinois

In numerous laboratory, greenhouse, and small- and large-scale field tests in the states of Florida and Louisiana, *Cercospora rodmanii* has shown good potential as a biocontrol agent for waterhyacinth (*Eichhornia crassipes*). This endemic fungus readily invades waterhyacinth plants by way of stomates on the leaves and petioles. It causes leaf-spotting with resultant death of severely infected leaves. Diseased plants show a general debilitation as newer leaves are killed. Eventually, entire plants succumb and gradually sink. In severe epiphytotics, an increasingly larger number of leaves and entire plants die, and open water becomes apparent in areas that in preceding weeks had been totally covered with waterhyacinths. An epiphytotic can be induced by spraying waterhyacinths with either freshly prepared or formulated inoculum of *C. rodmanii*. Once established, the fungus spreads by airborne conidia to plants in adjacent, nontreated areas.

CONTROL OF STRANGLER VINE IN CITRUS GROVES WITH *PHYTOPHTHORA PALMIVORA*: STUDIES ON CITRUS SUSCEPTIBILITY

W. H. RIDINGS, C. L. SCHOULTIES, M. E. KANNWISCHER, S. H. WOODHEAD, AND N. E. EL-GHOLL

Division of Plant Industry, Florida Department of Agriculture and Consumer Services, Gainesville, Florida

Plant Pathology Department, University of Florida, Gainesville, Florida

Chemical and Agricultural Products Division, Abbott Laboratories, North Chicago, Illinois

The strangler vine or milkweed vine (*Morrenia odorata*) is a major weed pest in many of Florida's citrus-growing areas. The vine competes with citrus trees for light, water, and nutrients; can girdle tree limbs; and interferes with spraying, harvesting, and irrigation practices. A Florida isolate of *Phytophthora palmivora* (formerly identified as *P. citrophthora*) has been very efficacious as a biological control agent of strangler vine. Greater than 90% of the vines in citrus groves treated with 20 chlamydospores of this isolate per square centimeter of soil died within 10 weeks when soil moisture and temperature were not limiting to disease development.

Studies were conducted to determine whether the *P. palmivora* isolate is a pathogen of citrus rootstocks, budded trees, or fruit. Pathogenicity of this isolate to citrus was compared with that of an isolate of *P. parasitica* from citrus. No disease was observed in any of eight common citrus rootstocks inoculated with *P. palmivora*, whereas inoculation with *P. parasitica* resulted in root rot on 80–100% of the plants of each rootstock. Oospores were not observed in the roots of plants inoculated with *P. palmivora* (A_2 mating type) and *P. parasitica* (A_1 mating type), and the amount of root rot was not greater than with *P. parasitica* alone. None of the budded trees died or declined in a test where 13 citrus rootstock-scion combinations were inoculated with *P. palmivora*. When *P. palmivora* was sprayed onto citrus fruit on trees, only 6% of the inoculum was viable within 1 hour of spraying,

and none of the fruit developed brown rot when placed in humidity chambers for
10 days.

The isolate of *P. palmivora* from strangler vine did not appear to be a pathogen
of citrus under the conditions studied.

BIOLOGICAL CONTROL OF PRICKLY SIDA, *SIDA SPINOSA*, WITH *COLLETOTRICHUM MALVARUM*

D. O. TeBeest, G. E. Templeton, and R. J. Smith, Jr.
Department of Plant Pathology, University of Arkansas, Fayetteville, Arkansas
U.S. Department of Agriculture, Science and Education Administration,
Agricultural Research, Stuttgart, Arkansas

Colletotrichum malvarum causes an anthracnose of prickly sida, *Sida spinosa*, a
weed common to several crops throughout the southeastern United States. Studies
conducted since 1975 in the field, laboratory, and greenhouse indicate that this
fungus has considerable potential as a biological control agent for prickly sida. The
fungus is specific for prickly sida, hollyhock (*Althaea rosea*), and several other
malvaceous hosts excluding cotton. Disease development is rapid and severe when
plants are sprayed with 2×10^6 spores/ml and held in a dew chamber for 24 hours
at 24°C, or when given split dew periods of 12–16 hours at 24°C or 12 hours at
20°C followed by 28°C. Control of prickly sida in the field has been erratic and
dependent on environmental conditions at the time of inoculum application. Best
control (90–95%) was achieved when inoculum was applied while cool (24°C),
moist conditions prevailed for several days following inoculation. Obstacles to the
commercial development and effective use of *C. malvarum* as a bioherbicide include (1) limited and inconsistent sporulation in liquid culture, (2) environmental
restrictions on infection and disease development, and (3) the necessity for multiple
inoculum applications to control the repeated emergence of new seedlings of
prickly sida.

BIOLOGICAL CONTROL OF NORTHERN JOINTVETCH, *AESCHYNOMENE VIRGINICA*, WITH *COLLETOTRICHUM GLOEOSPORIOIDES* F. SP. *AESCHYNOMENE*

G. E. Templeton, D. O. TeBeest, and R. J. Smith, Jr.
Department of Plant Pathology, University of Arkansas, Fayetteville, Arkansas
U.S. Department of Agriculture, Science and Education Administration,
Agricultural Research, Stuttgart, Arkansas

The fungus *Colletotrichum gloeosporioides* f. sp. *aeschynomene* (CGA) has been
used to control northern jointvetch, *Aeschynomene virginica*, in Arkansas rice and

soybean fields since 1969. From 1972 through 1980 field tests were conducted under experimental use permits issued by the U.S. Environmental Protection Agency (EPA) and the Arkansas State Plant Board. A total of 123 rice fields and 26 soybean fields consisting of 2005 ha were treated with fresh- or dry-spore formulations of the fungus. From 1976 through 1979 dry-spore formulations prepared by The Upjohn Company, Kalamazoo, MI, were applied to 375 ha of rice and averaged 87% control. From 1972 through 1979 fresh spores applied to 703 ha of rice yielded an average of 95% control. From 1976 through 1979 dry-spore formulations applied to 63 ha of soybean averaged 99.8% control. During this same period fresh spores applied to 60 ha averaged 99.6% control. Periods of cloudy, rainy weather within 4 weeks of inoculation enhanced disease development and weed control, but such weather was not required for the success of this biocontrol agent. *Colletotrichum gloeosporioides* f. sp. *aeschynomene* is specific to the genus *Aeschynomene* and infects *A. evenia, A. indica, A. pratensis, A. rudis, A. scabra, A. sensitiva,* and *A. virginica,* but not *A. americana, A. brasiliana, A. falcata, A. histrix, A. paniculata,* and *A. villosa.* Only *A. virginica* was killed by CGA.

BIOLOGICAL CONTROL OF SPURRED ANODA (*ANODA CRISTATA*) WITH *ALTERNARIA MACROSPORA*

H. L. WALKER

Southern Weed Science Laboratory, Agricultural Research, Science and Education Administration, U.S. Department of Agriculture, Stoneville, Mississippi

Alternaria macrospora, an indigenous pathogen, has been studied for the biological control of spurred anoda (*Anoda cristata*), an important weed in cotton growing areas of the southern United States. Host range studies indicate cotton (*Gossypium hirsutum* and *G. barbadense*), tomato (*Lycopersicon esculentum*), soybean (*Glycine max*), corn (*Zea mays*), rice (*Oryza sativa*), wheat (*Triticum aestivum*), and other representative crop and weed species to be resistant to the pathogen.

Methods have been developed to produce inoculum for greenhouse and field studies. Mycelium is grown in submerged liquid culture, comminuted, poured into pans, and exposed to alternating cool-white fluorescent light. The conidia are air-dried and harvested with a cyclone collector. These conidial preparations contain approximately 2×10^8 spores/g and retain 90% or higher viability for at least 1 year when stored at 4°C. Granular formulations are produced by mixing blended mycelium with vermiculite, clay particles, corn-cob-grits, or other suitable carriers before sporulation is induced.

Foliar applications containing 1 to 5×10^5 spores/ml reduced dry weights of spurred anoda seedlings 70–75% after 4–6 weeks in the greenhouse or field. Pre-emergence or postemergence applications of granular formulations of the pathogen reduced dry weights of spurred anoda seedlings 64–70% after 4 weeks in the greenhouse or field.

The major constraints to control of spurred anoda with this pathogen appear to be the duration of free moisture required for optimum disease development (24 h or longer) and an increase in tolerance to the pathogen as the weed increases in size.

UREDO EICHHORNIAE, A PATHOGEN OF EICHHORNIA CRASSIPES

R. Charudattan

Plant Pathology Department, University of Florida, Gainesville, Florida

Uredo eichhorniae is a rust pathogen of waterhyacinth (*Eichhornia crassipes*) found in the humid, semitropical regions of Argentina, Uruguay, and Brazil. In uredospore morphology, physiology, and pathogenicity to aquatic hosts, *Uredo eichhorniae* resembles two *Uromyces* species and two undescribed uredial rusts found on other members of the Pontederiaceae. However, *Uredo eichhorniae* does not infect pickerelweed (*Pontederia lanceolata*), and *Uromyces pontederiae*, a pathogen of pickerelweed, is nonpathogenic to waterhyacinth.

The uredospores of *Uredo eichhorniae* are the primary source of inoculum in nature, and to date the alternate host(s) and the missing spore stages of this fungus have not been found. The uredospores can be stored in liquid nitrogen for at least 18 months without apparent loss of viability or virulence. Uredospores of *Uredo eichhorniae* and *Uromyces pontederiae* do not germinate merely on contact with free water but can be stimulated to germinate readily *in vitro* and on host leaves with the aid of certain ketones and aldehydes.

Attempts to induce the formation of teliospores by *Uredo eichhorniae* on waterhyacinth have failed thus far; hence the life cycle of this fungus is not fully understood. More information on its life cycle and efficacy as a biological control agent must be available before *Uredo eichhorniae* could be considered for introduction into other areas of the world.

CONTROL OF CANADA THISTLE BY A RUST, PUCCINIA OBTEGENS

W. E. Dyer, S. K. Turner, P. K. Fay, E. L. Sharp, and D. C. Sands

Departments of Plant and Soil Science, and Plant Pathology, Montana State University, Bozeman, Montana

Canada thistle (*Cirsium arvense*) is a noxious, rhizomatous perennial weed that infests more than 3.7 million ha of cropland in Montana. *Puccinia obtegens*, an autoecious, endemic rust pathogen, is being investigated as a biocontrol agent for this weed. Host genetic resistance, length of dew period following spore inoculation, temperature, and host growth stage influence disease establishment. A stable, protein-based foam was used to create an artificial environment to overcome limitations from temperature and dew period length. A temperature-controlled spore releasing device was designed to provide inoculum during early host growth stages. This device has been useful in the spread of other rust pathogens in the field.

BIOLOGICAL CONTROL OF MORNINGGLORY SPECIES WITH A RUST, *COLEOSPORIUM IPOMOEAE*

D. O. TeBeest

Department of Plant Pathology, University of Arkansas, Fayetteville, Arkansas

A disease of morningglory species caused by the rust fungus *Coleosporium ipomoeae* has been investigated since 1978 for potential use of this fungus as a biological control agent. Annual morningglories such as *Ipomoea lacunosa* (small-white), *I. hederacea* (ivyleaf), *I. hederacea* var. *integriuscala* (entireleaf), *I. muricata* (purple moonflower), *I. wrightii* (willowleaf), and *Jacquemontia tamnifolia* (smallflower) are becoming the dominant weeds in Arkansas soybean fields. *Ipomoea hederacea, I. hederacea* var. *integriuscala, I. wrightii, I. lacunosa, I. nil,* and *I. coccinea* are susceptible to infection by uredospores of *C. ipomoeae*, whereas *I. hederifolia, I. muricata, I. quamoclit, I. pandurata, I. purpurea,* and *J. tamnifolia* are not infected. Uredospores, teliospores, and sporidia are produced on morningglory, whereas pycniospores and aeciospores are produced on the alternate hosts, *Pinus* species. In field tests conducted in 1979, epidemics initiated on May 31 on *I. hederacea* and *I. hederacea* var. *integriuscala* resulted respectively in 69 and 59% of leaf area rusted by August 19. These epidemics developed too slowly (r = 0.04 units/day) to prevent growth, competition with crops, and reseeding by each host species. These results indicate that reliance on sustained, self-propagating epidemics of *C. ipomoeae* to control morningglory in the field would not be practical.

POTENTIAL USE OF *PUCCINIA XANTHII* (RUST FUNGUS) FOR BIOCONTROL OF COCKLEBUR (*XANTHIUM STRUMARIUM*)

C. G. Van Dyke and H. W. Spurr, Jr.

Departments of Botany and Plant Pathology, North Carolina State University, Raleigh, North Carolina

Tobacco Research Laboratory, U.S. Department of Agriculture, Science and Education Administration, Agricultural Research, Oxford, North Carolina

Puccinia xanthii is a microcyclic, autoecious rust fungus that occurs worldwide on cocklebur (*Xanthium strumarium*). This fungus is being studied for use as a biocontrol agent; it is the only major pathogen consistently infecting cocklebur. More than 30 counties in North Carolina have been surveyed and found to have rust-diseased cocklebur. In some fields only one or two plants were diseased, whereas in other fields nearly every plant had rust pustules. Teliospores produced in the rust pustule germinate readily in a moist environment at room temperature when taken directly from fresh leaves, but if the leaf dries, the spores become dormant. Several successive freezing and thawing regimes seem necessary to break this dormancy. In the greenhouse, heavy infection and stunting of cocklebur seedlings inoculated with dormancy-broken teliospores have been achieved. However, field inoculations have not been successful. We are continuing greenhouse studies of fungal infection and disease development and plan to work closely with weed scientists in field studies.

ARAUJIA MOSAIC VIRUS, A POTENTIAL BIOLOGICAL CONTROL AGENT FOR MILKWEED VINES

R. Charudattan and D. S. Heron
Plant Pathology Department, University of Florida, Gainesville, Florida

Morrenia odorata and *Araujia sericofera* are South American milkweed vines (Asclepiadaceae) that have become troublesome weeds in citrus groves of Florida and California, respectively. The *Morrenia* problem in Florida is particularly serious. These vines spread mainly by comose seeds produced in dehiscent follicles; usually more than 500 seeds per follicle are produced and wind dispersed. Regeneration of the vines from roots and subterranean stems is another important means of proliferation.

A virus disease of *Morrenia* spp. and *Araujia* spp. is common throughout northern Argentina, the likely native home for these plants. The virus had the following characteristics, suggesting that it is a member of the potato virus Y (potyvirus) group: flexuous, 600–800 nm long rod-shaped virus particles; pinwheel, circular, and the "bundle-of-needles-like" cytoplasmic virus-inclusion bodies; and stylet-borne aphid transmissibility. On the basis of a literature review, host range, and biochemical and serologic data on hand, the virus has been considered to be a new member of the potyvirus group and named *Araujia mosaic virus* (AjMV).

This virus induces mild to severe mosaic symptoms, the latter accompanied occasionally by leaf necrosis. Infected plants become stunted and may eventually die. The virus induces seedling mortality in certain milkweed vine species. The virus is systemic in the vines. A general reduction in the biomass of infected plants results.

Araujia mosaic virus has been tested on 121 plant species in 25 families through manual inoculations; 37 of these species were also tested through aphid inoculations. Only 10 species, all members of Asclepiadaceae and vines, were susceptible. Herbaceous milkweeds and plants outside the milkweed family were nonsusceptible. The highly restricted host range and the ability to stunt the vines are features that indicate that AjMV is a safe and desirable biological control agent for milkweed vines.

BIOCONTROL OF RUSSIAN KNAPWEED WITH A NEMATODE

A. K. WATSON

Department of Plant Science, Macdonald Campus of McGill University, Sainte Anne De Bellevue, Quebec, Canada

Russian knapweed (*Acroptilon repens* = *Centaurea repens*) is a troublesome, introduced perennial weed that is widely distributed in western Canada and in the western and central regions of the United States. In the native range of the weed, a nematode, *Paranguina picridis*, is being utilized as a biological control agent and is being artificially spread by spraying suspensions containing the nematode. The nematode was investigated as a possible biocontrol agent of Russian knapweed in North America. The laboratory host range of the nematode was restricted to a few species of the Centaureinae and Carduinae subtribes of the Cynareae tribe of the Asteraceae family, but Russian knapweed was the only plant species rated as susceptible to *P. picridis* attack. Plant galls on Russian knapweed developed extensive layers of nutritive tissue with minimal cell necrosis. Galls on other test species did not develop extensive feeding sites, and extensive cell necrosis was commonly observed.

PLANTS CITED

Family, Genus, and Species	Common Name
Amaranthaceae	
Alternanthera	
philoxeroides (Mart.) Griseb.	alligatorweed
Amaranthus	
retroflexus L.	redroot pigweed
Apocynaceae	
Apocynum	
cannabinum L.	hemp dogbane
Araceae	
Pistia	
stratiotes L.	waterlettuce
Asclepiadaceae	
Araujia	
sericofera Brotero	bladder flower
Asclepias	
syriaca L.	common milkweed
Morrenia	
odorata (Hook. and Arn.) Lindley	strangler vine (milkweed vine)
Sarcostemma	
cynanchoides Dcne.	climbing milkweed
Bignoniaceae	
Campsis	
radicans (L.) Seem.	trumpetcreeper

Reference sources: *Weed Sci.* **19**, 435–476, 1971; H. K. Airy Shaw. 1973. *A Dictionary of the Flowering Plants and Ferns*, 8th ed. Cambridge University Press, Cambridge, England, 1245 pp; M. L. Fernald. 1970. *Gray's Manual of Botany*, 8th ed. Van Nostrand, New York, 1632 pp; and others.

Family, Genus, and Species	Common Name
Cactaceae	
Opuntia spp.	pricklypears
Cannabinaceae	
Cannabis	
sativa L.	hemp
Caryophyllaceae	
Silene	
alba Muhl.	white campion
dioica (L.) Clairv.	red campion
Chenopodiaceae	
Beta	
vulgaris L.	sugarbeet
Chenopodium	
album L.	common lambsquarters
strictum Roth	goosefoot
other species	
Commelinaceae	
Commelina	
diffusa Burm. f.	spreading dayflower
Compositae	
Acanthospermum	
hispidum DC.	bristly starbur
Acroptilon	
repens (L.) DC.	
(= *Centaurea repens* L.)	Russian knapweed
Ageratina	
riparia (Regel) King and	river eupatorium
Robins.	(hamakua pamakani)
(= *Eupatorium riparium*	
Regel)	
Ambrosia	
artemisiifolia L.	common ragweed
other species	
Carduus	
acanthoides L.	plumeless thistle
nutans L.	musk thistle
Centaurea	
solstitialis L.	yellow starthistle

Family, Genus, and Species	Common Name
Chondrilla	
juncea L.	rush skeletonweed (skeletonweed)
other species	
Cirsium	
arvense (L.) Scop.	Canada thistle
Eclipta	
alba (L.) Hassk.	yerba-de-tago (eclipta)
Eupatorium	
adenophorum Spr.	pamakani (Crofton weed)
riparium Regel	hamakua pamakani
Helianthus	
annuus L.	sunflower
ciliaris DC.	Texas blueweed
tuberosus L.	Jerusalem artichoke
other species	
Lactuca	
serriola L.	prickly lettuce
Senecio	
vulgaris L.	common groundsel
Silybum	
marianum (L.) Gaertn.	milk thistle
Sonchus	
arvensis L.	perennial sowthistle
Taraxacum	
officinale Weber	common dandelion
Tragopogon	
dubius Scop.	western salsify
mirus Ownbey	goatsbeard
miscellus Ownbey	goatsbeard
porrifolius L.	common salsify
pratensis L.	meadow salsify
other species	
Xanthium	
canadense Mill.	cocklebur
pensylvanicum Wallr.	common cocklebur
pungens Wallr.	Noogoora burr

Family, Genus, and Species	Common Name
spinosum L.	spiny cocklebur
strumarium L.	heartleaf cocklebur
Convolvulaceae	
Convolvulus	
arvensis L.	field bindweed
sepium L.	hedge bindweed
Cuscuta	
campestris Yunck.	field dodder
cupulata Engelm.	dodder
epithymum Murr.	dodder
Ipomoea	
coccinia L.	scarlet morningglory
hederacea (L.) Jacq.	ivyleaf morningglory
hederacea (L.) Jacq. var.	
integriuscala Gray	entireleaf morningglory
hederifolia L.	scarlet starglory
lacunosa L.	small-white morning-glory
muricata (L.) Jacq.	purple moonflower
nil (L.) Roth	white-edge morningglory
pandurata (L.) G.F.W. Mey	bigroot morningglory
purpurea (L.) Roth	tall morningglory
quamoclit L.	cypressvine morningglory
wrightii Gray	willowleaf morningglory
other species	
Jacquemontia	
tamnifolia (L.) Griseb.	smallflower morning-glory
Cruciferae	
Brassica	
campestris L.	wild turnip
Capsella	
bursa-pastoris (L.) Medic.	shepherdspurse
Raphanus	
raphanistrum L.	wild radish
sativus L.	radish
Cyperaceae	
Cyperus	
esculentus L.	yellow nutsedge (yellow nutgrass)

Family, Genus, and Species	Common Name
rotundus L.	purple nutsedge (purple nutgrass)
other species	
Dennstaedtiaceae	
Pteridium	
aquilinum (L.) Kuhn	bracken fern
Ebenaceae	
Diospyros	
virginiana L.	persimmon
Euphorbiaceae	
Croton spp.	crotons
Euphorbia	
cyparissias L.	cypress spurge
esula L.	leafy spurge
other species	
Fagaceae	
Castanea	
dentata (Marsh.) Borkh.	American chestnut
mollissima Blume	Chinese chestnut
Quercus	
alba L.	white oak
rubra L.	northern red oak
virginiana Mill.	live oak
Geraniaceae	
Erodium	
cicutarium (L.) L'Hér.	redstem fillaree
Gramineae	
Aegilops	
cylindrica Host	jointed goatgrass
Agropyron	
repens (L.) Beauv.	quackgrass
Avena	
barbata Brot.	slender oat
fatua L.	wild oat
sativa L.	oat
Brachiaria	
ciliatissima (Buckl.) Chase	fringed signalgrass
platyphylla (Griseb.) Nash	broadleaf signalgrass

Family, Genus, and Species	Common Name
Bromus	
commutatus Schrad.	hairy chess
japonicus Thunb.	Japanese brome
mollis L.	soft chess
secalinus L.	cheat
other species	
Cortaderia	
jubata (Lem.) Stapf.	pampas grass
Cynodon	
dactylon (L.) Pers.	bermudagrass
Digitaria spp.	crabgrasses
Echinochloa	
crus-galli (L.) Beauv.	barnyardgrass
Eleusine	
coracana (L.) Gaertn.	finger millet
indica (L.) Gaertn.	goosegrass
Eragrostis	
curvula (Schrad.) Nees	weeping lovegrass
Festuca	
rubra L.	red fescue
Holcus	
mollis L.	German velvetgrass
Hordeum	
murinum L.	wall barley
Imperata	
cylindrica (L.) Beauv.	cogongrass
Leptochloa	
fascicularis (Lam.) Gray	bearded sprangletop
Lolium spp.	darnels (ryegrasses)
Muhlenbergia	
frondosa (Poir.) Fern.	wirestem muhly
Oryza	
sativa L.	rice
Panicum	
dichotomiflorum Michx.	fall panicum
maximum Jacq.	guineagrass
texanum Buckl.	Texas panicum
Poa	
annua L.	annual bluegrass

Family, Genus, and Species	Common Name
Rottboellia	
exaltata L. f.	itchgrass
Saccharum	
officinarum L.	sugarcane
Secale	
cereale L.	rye
Setaria	
faberi Herrm.	giant foxtail
Sorghum	
bicolor (L.) Moench	sorghum
halepense (L.) Pers.	johnsongrass
Spartina	
anglica C. E. Hubbard	cordgrass
Triticum	
aestivum L.	wheat
Zea	
mays L.	maize (corn)
Guttiferae	
Hypericum	
perforatum L.	St. Johnswort
Haloragaceae	
Myriophyllum	
brasiliense Camb.	parrotfeather
spicatum L.	eurasian watermilfoil
Hydrocharitaceae	
Elodea	
canadensis Michx.	elodea
Hydrilla	
verticillata (L. f.) Royle	hydrilla
Labiatae	
Galeopsis	
tetrahit L.	hempnettle
Lamium	
moluccellifolium Fries	deadnettle
Moluccella sp. (= *Molucella*)	molluccabalm
Lauraceae	
Persea spp.	avocados

Family, Genus, and Species	Common Name

Leguminosae
  Aeschynomene

americana L.	American jointvetch
brasiliana (Poir.) DC.	Brazilian jointvetch
evenia Wright in Saur.	jointvetch
falcata (Poir.) DC.	jointvetch
histrix Poir. in Lam.	jointvetch
indica L.	Indian jointvetch
paniculata Willd. ex Vog.	jointvetch
pratensis Small	jointvetch
rudis Benth.	jointvetch
scabra G. Don	jointvetch
sensitiva Sw.	sensitive jointvetch
villosa Poir.	jointvetch
virginica (L.) B. S. P.	northern jointvetch

  Albizzia

julibrissin Durazzini	silktree albizzia

  Arachis

hypogaea L.	peanut

  Cassia

obtusifolia L.	sicklepod
surrattensis Lamarck	brushweed

  Glycine

max (L.) Merr.	soybean

  Medicago

sativa L.	alfalfa

  Sesbania

exaltata (Raf.) Cory	hemp sesbania

Liliaceae
  Allium

canadense L.	wild onion
vineale L.	wild garlic

Loranthaceae
  Arceuthobium spp.   dwarf mistletoes

Lythraceae
  Ammannia

coccinea Rottb.	purple ammannia
other species	

Family, Genus, and Species	Common Name
Malvaceae	
Abutilon	
theophrasti Medic.	velvetleaf
Althaea	
rosea L.	hollyhock
Anoda	
cristata (L.) Schlecht.	spurred anoda
Gossypium	
barbadense L.	sea island cotton
hirsutum L.	cotton
Sida	
spinosa L.	prickly sida
Menyanthaceae	
Nymphoides	
orbiculata (Sm.) Ktze.	waterlily
Myrtaceae	
Eucalyptus	
marginata Sm.	jarrah (gumtree)
Nymphaeaceae	
Brasenia	
schreberi Gmel.	watershield
Nuphar	
luteum (L.) Sm.	yellow waterlily
Nymphaea	
odorata Ait.	fragrant waterlily
tuberosa Paine	white waterlily
Onagraceae	
Jussiaea	
decurrens (Walt.) DC.	winged waterprimrose
Oenothera	
biennis L.	common eveningprimrose
Orobanchaceae	
Orobanche spp.	broomrapes
Oxalidaceae	
Oxalis	
corniculata L.	creeping woodsorrel
pes-caprae L.	Bermuda buttercup

Family, Genus, and Species	Common Name

Pinaceae
  Pinus
    echinata Mill. — shortleaf pine
    taeda L. — loblolly pine
    other species

Polygonaceae
  Emex
    australis Steinh. — emex
    spinosa Campd. — spiny emex
  Polygonum spp. — smartweeds
  Rumex
    crispus L. — curly dock
    other species

Pontederiaceae
  Eichhornia
    azurea (Swartz) Kunth. — anchored waterhyacinth
    crassipes (Mart.) Solms — waterhyacinth
  Heteranthera
    limosa (Sw.) Willd. — ducksalad
  Monochoria
    vaginalis Presl. — monochoria
  Pontederia
    lanceolata Nutt. — pickerelweed

Portulacaceae
  Portulaca
    oleracea L. — common purslane

Rosaceae
  Malus spp. — apples
  Rubus
    constrictus Lef. and M. — wild blackberry
    penetrans Bailey — Hawaiian blackberry
    ulmifolius Schott — wild blackberry

Rubiaceae
  Coffea spp. — coffees

Rutaceae
  Citrus spp. — citruses

Family, Genus, and Species	Common Name
Salicaceae	
Populus	
alba L.	white poplar
other species	
Salix spp.	willows
Salviniaceae	
Salvinia	
molesta Mitchell	giant salvinia
other species	
Scrophulariaceae	
Bacopa	
rotundifolia (Michx.) Wettst.	waterhyssop
Verbascum	
thapsus L.	common mullein
Solanaceae	
Datura	
stramonium L.	jimsonweed
Lycopersicon	
esculentum Mill.	tomato
pimpinellifolium (Jusl.) Mill.	currant tomato
Nicotiana	
tabacum L.	tobacco
Physalis spp.	groundcherries
Solanum	
carolinense L.	horsenettle
dulcamara L.	bitter nightshade
elaeagnifolium Cav.	silverleaf nightshade
rostratum Dunal	buffalobur
tuberosum L.	potato
other species	
Tamaricaceae	
Tamarix spp.	saltcedars (tamarisks)
Theaceae	
Camellia	
sinensis (L.) Kuntze	tea

257

Family, Genus, and Species	Common Name
Turneraceae	
Turnera	
ulmifolia L.	yellow alder
Typhaceae	
Typha	
domingensis Pers.	southern cattail
latifolia L.	common cattail
Ulmaceae	
Ulmus	
americana L.	American elm
parvifolia Jacq.	Chinese elm
Umbelliferae	
Hydrocotyle	
umbellata L.	water pennywort

MICROORGANISMS, VIRUSES, PHAGES, NEMATODES, AND ARTHROPODS CITED

MICROORGANISMS

FUNGI

Division: Eumycota
  Subdivision: Mastigomycotina
    Class: Oomycetes
      Order: Peronosporales
        Albugo tragopogonis Pers.
        Phytophthora cinnamomi Rands
        P. citrophthora (R. E. Sm. and E. H. Sm.) Leonian
        P. infestans (Montagne) de Bary
        P. palmivora (Butler) Butler
        P. parasitica Dast.
        Pythium sp.

  Subdivision: Zygomycotina
    Class: Zygomycetes
      Order: Entomophthorales
        Entomophthora virulenta Hall and Dunn

Based on taxonomic treatments in G. C. Ainsworth. 1971. *Ainsworth and Bisby's Dictionary of the Fungi*, 6th ed. Commonwealth Mycological Society, Kew, England, 663 pp.; G. C. Ainsworth, F. K. Sparrow, and A. S. Sussman. 1973. *The Fungi: An Advanced Treatise*. Vols. 4A and 4B. Academic Press, New York, 621 and 504 pp.; J. G. Holt, ed. 1977. *The Shorter Bergey's Manual of Determinative Bacteriology*, 8th ed. Williams & Wilkins, Baltimore, MD, 356 pp.; and others.

Subdivision: Ascomycotina
  Class: Plectomycetes
    Order: Microascales
      Ceratocystis fagacearum (Bretz) Hunt
      C. ulmi (Buisman) C. Moreau
      Ceratostomella ulmi Buism. (see *Ceratocystis ulmi*)
    Order: Erysiphales
      Erysiphe cichoracearum DC. ex Merat
      Leveillula taurica (Lév.) Arnaud
  Class: Pyrenomycetes
    Order: Hypocreales
      Nectria fuckeliana Booth var. *macrospora*
    Order: Clavicipitales
      Claviceps purpurea (Fr.) Tul.
    Order: Sphaeriales
      Cochliobolus carbonum Nelson
      C. heterostrophus Drechs.
      C. species
      Endothia parasitica (Murrill) Anderson and Anderson
      Glomerella sp.
      Hypoxylon spp.
      Setosphaeria turcica (Luttrell) Leonard and Sugg.
      Valsa spp.
      Wallrothiella arceuthobii (Peck) Sacc.
  Class: Discomycetes
    Order: Helotiales
      Monilinia sp.
      Sclerotinia fructicola Winter
      S. homoeocarpa Bennett
  Class: Loculoascomycetes
    Order: Pleosporales
      Botryosphaeria tamaricis (Cke.) Th. and Syd.
      Venturia inaequalis (Cooke) Aderhold
Subdivision: Basidiomycotina
  Class: Teliomycetes
    Order: Uredinales
      Aecidium asclepiadinum Speg.
      Coleosporium ipomoeae (Schw.) Burr
      Hemileia vastatrix Berk. and Br.
      Melampsora lini (Ehrenb.) Lév.
      Phragmidium violaceum (Schultz) Wint.
      Puccinia acroptili Syd.
      P. araujae Lév.

P. canaliculata (Schw.) Lagh.

P. chondrillina Bubak and Syd.

P. coronata Corda

P. graminis Pers.

P. jaceae Otth

P. obtegens (Link) Tul.

P. oxalidis (Lév.) Diet. and Ellis

P. sorghi Schw.

P. species

P. xanthii Schw.

Uredo eichhorniae Gonz.-Frag. and Cif.

Uromyces pontederiae Gerard

U. rumicis (Schum.) Wint.

Order: Ustilaginales

Sorosporium cenchri Henn.

Sphacelotheca cruenta (Kuehn) Potter

Tilletia spp.

Ustilago hordei (Pers.) Lagerh.

U. violacea (Pers.) Roussel

Class: Hymenomycetes

Order: Agaricales

Armillariella (*Armillaria*) *mellea* (Vahl ex Fries) Kummer

Subdivision: Deuteromycotina

Class: Hyphomycetes

Acremonium zonatum (Saw.) Gams

Alternaria alternantherae Holcomb and Antonopoulos

A. cuscutacidae Rudakov

A. eichhorniae Nag Raj and Ponnappa

A. macrospora Zimm.

A. species

Aspergillus nidulans Eidam

A. oryzae (Ahlburg) Cohn

A. species

Bipolaris stenospila Drechs.

B. species

Cephalosporium diospyri Crandell

C. species

Cercospora eupatorii Peck

C. hydrocotyles Ellis and Everh.

C. molucellae Bremer and Petrak

C. nymphaeacea Cooke and Ellis

C. piaropi Tharp.

C. rodmanii Conway

C. species
Cercosporella ageratinae (nomen nudem)
Chalara guercina Henry
Curvularia sp.
Cylindrocladium crotalariae Loos
Dichotomophothoropsis nymphaearum (Rand) M. B. Ellis
Exserohilum sp.
Fusarium oxysporum Schlecht. f. sp. *cannabis* Snyder and Hans.
F. oxysporum Schlecht. var. *orthoceras* (Appel and Wollenw.) Bilay
F. oxysporum Schlecht. f. sp. *perniciosum* (Hept.) Toole
F. roseum (Link) Snyder and Hans.
F. roseum 'Culmorum' (Link ex Fr.) Snyder and Hans.
F. solani (Mart.) Appel and Wollenw.
F. species
Helminthosporium spp.
Hirsutella thompsonii Fisher
Myrothecium roridum Tode ex Fr.
Penicillium chrysogenum Thom
P. species
Pyricularia sp.
Rhizoctonia sp.
Sclerotium rolfsii (Sacc.) Curzi
Spicariopsis sp.
Trichoderma sp.
Class: Coelomycetes
 Order: Melanconiales
  Colletotrichum destructivum O'Gara
  C. gloeosporioides (Penz.) Sacc. f. sp. *aeschynomene*
  C. gloeosporioides (Penz.) Sacc. f. sp. *jussiaeae*
  C. lindemuthianum (Sacc. and Magn.) Briosi and Cav.
  C. malvarum (A. Braun and Casp.) Southworth
  C. phomoides (Sacc.) Chester
  C. species
 Order: Sphaeropsidales
  Phyllosticta maydis Arny and R. R. Nelson
  P. species

BACTERIA

Class: Schizomycetes
 Order: Pseudomonadales
  Bdellovibrio bacteriovorus Stolp and Starr

Pseudomonas solanacearum E. F. Sm.
P. syringae van Hall
Order: Eubacteriales
Agrobacterium radiobacter Beijerinck and van Delden
Bacillus lentimorbus Dutky
B. popilliae Dutky
B. thuringiensis Berliner
B. thuringiensis Berliner var. *aizawai*

PROTOZOA

Nosema locustae Canning

VIRUSES AND PHAGES

Araujia mosaic virus
Cyanophages
Nuclear polyhedrosis virus (NPV)
Douglas fir tussock moth NPV
Gypsy moth NPV
Heliothis NPV
Tobacco mosaic virus

NEMATODES

Class: Phasmida
Order: Tylenchida
Aphelenchoides fragariae (Ritz.-Boz) Christie
Nothanguina phyllobia (Thorne) Thorne
Paranguina picridis Kirjanova and Ivanova

ARTHROPODS

Class: Arachnida
Order: Acarina
Orthogalumna terebrantis Wallwork (waterhyacinth mite)
Class: Insecta
Order: Orthoptera
Grasshoppers

Order: Hemiptera
  Blissus leucopterus (Say) (chinch bug)
  Oebalus pugnax (Fabricius) (rice stink bug)
Order: Homoptera
  Icerya purchasi Mask. (cottony cushion scale)
Order: Coleoptera
  Agasicles hygrophila Selman and Vogt (alligatorweed flea beetle)
  Chrysolina quadrigemina (Suffrian)
  Hylurgopinus rufipes (Eichhof) (American elm bark beetle)
  Lissorhoptrus oryzophilus Kuschel (rice water weevil)
  Neochetina eichhorniae Warner (mottled waterhyacinth weevil)
  Popillia japonica Newman (Japanese beetle)
  Rhinocyllus conicus (Froelich) (musk thistle weevil)
  Rodolia cardinalis (Muls.) (vedalia beetle)
  Scòlytus multistriátus (Marsham) (European elm bark beetle)
Order: Lepidoptera
  Bactra verutana Zell.
  Cactoblastis cactorum (Berg)
  Cucullia verbasci L. (mullein moth)
  Galleria mellonella (L.) (wax moth)
  Heliothis zea (Boddie) (bollworm)
  Hyles euphorbiae (L.) (spurge hawkmoth)
  Lymantria dispar (L.) (gypsy moth)
  Orgyia pseudotsugata (McDunnough) (Douglas fir tussock moth)
  Spodoptera frugiperda (J. E. Smith) (fall army worm)
  Vogtia malloi Pastrana

PESTICIDES CITED

Common Name	Chemical Name
Fungicides	
Benomyl	Methyl 1-(butylcarbamoyl)-2-benzimidazolecarbamate
Captafol	*cis*-N-[(1,1,2,2-Tetrachloroethyl)thio]-4-cyclohexene-1,2-dicarboximide
Captan	*cis*-N-[(Trichloromethyl)thio]-4-cyclohexene-1,2-dicarboximide
Carboxin	5,6-Dihydro-2-methyl-N-phenyl-1,4-oxathiin-3-carboxamide
Chloroneb	1,4-Dichloro-2,5-dimethoxybenzene
Chlorothalonil	Tetrachloroisophthalonitrile
Maneb	Manganese ethylenebisdithiocarbamate
PCNB	Pentachloronitrobenzene
PCNB-ETMT	Mixture of pentachloronitrobenzene + 5-ethoxy-3-trichloromethyl-1,2,4-thiadiazole
TCMTB	2-(Thiocyanomethylthio) benzothiazole
Thiabendazole	2-(4-Thiazolyl)-benzimidazole
Thiram	Bis(dimethylthio-carbamoyl)disulfide
Tricyclazole	5-Methyl-1,2,4-triazolo[3,4-*b*]-benzothiazole
Zinc ion-maneb complex	Coordination products of zinc ion and manganese ethylenebisdithiocarbamate
Herbicides	
Alachlor	2-Chloro-2′,6′-diethyl-N-(methoxymethyl)acetanilide

Common Name	Chemical Name

Herbicides

Atrazine	2-Chloro-4-ethylamino-6-isopropylamino-s-triazine
Benefin	N-Butyl-N-ethyl-α,α,α-trifluoro-2,6-dinitro-p-toluidine
Bentazon	3-Isopropyl-1H-2,1,3-benzothiadiazin-4(3H)-one 2,2-dioxide
Bifenox	Methyl 5-(2,4-dichlorophenoxy)-2-nitrobenzoate
Butachlor	N-(Butoxymethyl)-2-chloro-2′,6′-diethylacetanilide
Butylate	S-Ethyl diisobutylthiocarbamate
Cyanazine	2-[[4-Chloro-6-(ethylamino)-s-triazin-2-yl] amino]-2-methylpropionitrile
2,4-D	(2,4-Dichlorophenoxy)acetic acid
2,4-DB	4-(2,4-Dichlorophenoxy)butanoic acid
Dalapon	2,2-Dichloropropionic acid
Dinoseb	2-sec-Butyl-4,6-dinitrophenol
Diuron	3-(3,4-Dichlorophenyl)-1,1-dimethylurea
DSMA	Disodium methanearsonate
Fluchloralin	N-(2-Chloroethyl)-α,α,α-trifluoro-2,6-dinitro-N-propyl-p-toluidine
Fluometuron	1,1-Dimethyl-3-(α,α,α-trifluoro-m-tolyl) urea
Glyphosate	N-(Phosphonomethyl)glycine
Linuron	3-(3,4-Dichlorophenyl)-1-methoxy-1-methylurea
MCPA	[(4-Chloro-o-tolyl)oxy]acetic acid
Metolachlor	2-Chloro-N-(2-ethyl-6-methylphenyl)-N-(2-methoxy-1-methylethyl)acetamide
Metribuzin	4-Amino-6-(1,1-dimethylethyl)-3-(methylthio)-1,2,4-triazin-5(4H)-one
Molinate	S-Ethyl hexahydro-1H-azepine-1-carbothioate
MSMA	Monosodium methanearsonate
Naptalam	N-1-Naphthylphthalamic acid
Norflurazon	4-Chloro-5-(methylamino)-2-(α,α,α-trifluoro-m-tolyl)-3-(2H)-pyridazinone
Oryzalin	3,5-Dinitro N^4,N^4-dipropylsulfanilamide

266

Common Name	Chemical Name

Herbicides

Oxadiazon	2-*tert*-Butyl-4-(2,4-dichloro-5-isopropoxy-phenyl)-Δ^2-1,3,4-oxadiazolin-5-one
Paraquat	1,1'-Dimethyl-4,4'-bipyridinium ion
Pendimethalin	N-(1-Ethylpropyl)-3,4 dimethyl-2,6 dini-trobenzenamine
Profluralin	N-(Cyclopropylmethyl)-α,α,α-trifluoro-2, 6-dinitro-N-propyl-*p*-toluidine
Prometryn	2,4-Bis(isopropylamino)-6-methylthio-*s*-triazine
Propanil	3',4'-Dichloropropionanilide
Silvex	2-(2,4,5-Trichlorophenoxy)propionic acid
Simazine	2-Chloro-4,6-bis(ethylamino)-*s*-triazine
2,4,5-T	(2,4,5-Trichlorophenoxy)acetic acid
Thiobencarb	S-(4-Chlorobenzyl) N,N-diethylthiolcarba-mate
Trifluralin	α,α,α-Trifluoro-2,6-dinitro-N,N-dipropyl-*p*-toluidine
Vernolate	S-Propyl dipropylthiocarbamate

Insecticides

Acephate	O,S-Dimethyl acetylphosphoramidothioate
Aldicarb	2-Methyl-2-(methylthio) propionaldehyde O,(methylcarbamoyl)oxime
Azinphos-methyl	O,O-Dimethyl S-[(4-oxo-1,2,3-benzo-triazin-3(4H)-yl)methyl] phosphorodi-thioate
Bufencarb	m-(1-Methylbutyl)phenyl methylcarba-mate + m-(ethylpropyl)phenyl methyl-carbamate (3:1)
Carbaryl	1-Naphthyl N-methylcarbamate
Carbofuran	2,3-Dihydro-2,2-dimethyl-7-benzofuranyl methylcarbamate
Chlordimeform	N'-(4-Chloro-*o*-tolyl)-N,N-dimethylform-amidine
Chlorpyrifos	O,O-Diethyl O-(3,5,6-trichloro-2-pyridyl)-phosphorothioate
Dicrotophos	Dimethyl *cis*-2-dimethyl-carbamoyl-1-methylvinyl phosphate

Common Name	Chemical Name
	Insecticides
Dimethoate	*O,O*-Dimethyl *S*-(*N*-methylcarbamoyl-methyl) phosphorodithioate
Disulfoton	*O,O*-Diethyl *S*-[2-(ethylthio)ethyl]phosphorodithioate
Endosulfan	6,7,8,9,10,10-Hexachloro-1,5,5a,6,9,9a-hexahydro-6,9-methano-2,4,3-benzo(e)-dioxathiepin-3-oxide
Endrin	1,2,3,4,10,10-Hexachloro-6,7-epoxy-1,4,4a,5,6,7,8,8a-octahydro-1,4-*endo-endo*,8-dimethanonaphthalene
EPN	*O*-Ethyl *O*-(4-nitrophenyl) phenylphosphonothioate
Fenvalerate	Cyano(3-phenoxyphenyl)methyl 4-chloro-*a*-(1-methylethyl) benzeneacetate
Isofenphos	1-Methylethyl 2-[[ethoxy[(1-methylethyl)amino]phosphinothioyl]oxy]benzoate
Malathion	*O,O*-Dimethyl phosphorodithioate of diethyl mercaptosuccinate
Methamidophos	*O,S*-Dimethyl phosphoramidothioate
Methidathion	*O,O*-Dimethyl phosphorodithioate *S*-ester with 4-(mercaptomethyl)-2-methoxy-Δ^2-1,3,4-thiadiazolin-5-one
Methomyl	*S*-Methyl *N*-[(methylcarbamoyl)oxy]thioacetimidate
Methyl parathion	*O,O*-Dimethyl *O-p*-nitrophenyl phosphorothioate
Monocrotophos	Dimethyl phosphate ester with (*E*)-3-hydroxy-*N*-methyl-*cis*-crotonamide
Permethrin	3-(Phenoxyphenyl)methyl (±)-*cis,trans*-3-(2,2-dichloroethenyl)-2,2-dimethyl cyclopropanecarboxylate
Phorate	*O,O*-Diethyl *S*-[(ethylthio)methyl] phosphorodithioate
Propargite	2-[4-(1,1-Dimethylethyl)phenoxy]-cyclohexyl-2-propynyl sulfite
Sulprofos	*O*-Ethyl *O*-[4-(methylthio)phenyl]*S*-propyl phosphorodithioate
Toxaphene	Chlorinated camphene

Common Name	Chemical Name
	Insecticides
Trichlorfon	Dimethyl (2,2,2-trichloro-1-hydroxyethyl) phosphonate
	Nematicides
Aldicarb	See under Insecticides
D-D mixture	A 100% mixture of 1,3-dichloropropene, 1,2-dichloropropane, 3,3-dichloropropene, 2,3-dichloropropene, and other related chlorinated hydrocarbons
Ethoprop	O-Ethyl S,S-dipropyl phosphorodithioate
Metham	Sodium N-methyldithiocarbamate
Phenamiphos	Ethyl 3-methyl-4-(methylthio)phenyl (1-methylethyl) phosphoramidate

AUTHOR INDEX

SUBJECT INDEX

Abbott Laboratories, 38
Acephate, 194
Acremonium zonatum, 31, 192
Acroptilon repens, 30, 34, 246. *See also*
 Russian knapweed
Aecidium asclepiadinum, 33
Aeschynomene spp., 242
 americana, 242
 brasiliana, 242
 evenia, 242
 falcata, 242
 histrix, 242
 indica, 242
 paniculata, 242
 pratensis, 242
 rudis, 242
 scabra, 242
 sensitiva, 242
 villosa, 242
 virginica, 30, 139, 159, 223, 242. *See also*
  Northern jointvetch
Agasicles hygrophila, 191
Ageratina riparia, 30
Agrobacterium radiobacter, 158
Agroecosystem, 74, 89, 92, 116, 117, 120,
 190, 191
Agropyron repens, 222, 235. *See also*
 Quackgrass
Alachlor, 14, 194
Albizzia julibrissin, 30
Albugo tragopogonis, 30
Aldicarb, 194
Alfalfa, 8, 15, 17
Algae (blue-green), 30, 34
Allelopathic compounds, 24, 221
Alligatorweed, 3, 30, 38, 191, 225, 239. *See*
 also Alternanthera philoxeroides

Allozyme, 83–85
Alternanthera philoxeroides, 30, 91, 225, 239.
 See also Alligatorweed
Alternaria, spp., 31, 33
 alternantherae, 30, 239
 cuscutacidae, 31
 eichhorniae, 31
 macrospora (AM), 30, 151, 192, 199, 231,
  242
Althaea rosea, 241
Amaranthus retroflexus, 90
Ambrosia artemisiifolia, 30, 90
American elm bark beetle, 52
American Phytopathological Society, 114, 234
Ammannia coccinea, 75, 83
Androceras, section, 85
Anoda cristata, 30, 231, 242. *See also*
 Spurred anoda
Anthracnose, 241. *See also Colletotrichum* sp.
Aphelenchoides fragariae, 34
Apomict, 225
Apomixis, 82
Apple scab, 103
Arachis hypogaea, 106. *See also* Peanut
Araujia mosaic virus, 34, 245
Araujia sericofera, 245
Arceuthobium spp., 30
Armillariella mellea, 65
Arthropod(s), as biocontrol agents, 158, 176,
 191, 192, 200. *See also specific*
 arthropods
Asclepiadaceae, 245
Aspen, 33
Aspergillus sp., 32, 238
 nidulans, penicillin production, 108
 oryzae, 145
Atrazine, 9, 14, 90

279